建筑与都市系列丛书 | 世界建筑
Architecture and Urbanism Series | World Architecture

文筑国际 编译
Edited by CA-GROUP

Architecture in Norway and Denmark: Design with Nature

挪威与丹麦：自然设计

中国建筑工业出版社

图书在版编目（CIP）数据

挪威与丹麦：自然设计 = Architecture in Norway and Denmark Design with Nature : 汉英对照 / 文筑国际 CA-GROUP 编译 . -- 北京：中国建筑工业出版社，2021.1

（建筑与都市系列丛书 . 世界建筑）

ISBN 978-7-112-25722-5

Ⅰ . ①挪 … Ⅱ . ①文 … Ⅲ . ①建筑艺术 - 介绍 - 挪威 - 汉、英②建筑艺术 - 介绍 - 丹麦 - 汉、英 Ⅳ . ① TU-865.33 ② TU-865.34

中国版本图书馆 CIP 数据核字 (2020) 第 247367 号

责任编辑：毕凤鸣 刘文昕
版式设计：文筑国际
责任校对：王 烨

建筑与都市系列丛书 | 世界建筑
Architecture and Urbanism Series | World Architecture
挪威与丹麦：自然设计
Architecture in Norway and Denmark: Design with Nature
文筑国际 编译
Edited by CA-GROUP
*
中国建筑工业出版社出版、发行（北京海淀三里河路 9 号）
各地新华书店、建筑书店经销
北京雅昌艺术印刷有限公司 制版、印刷
*
开本：787 毫米 ×1092 毫米 1/16 印张：14 ½ 字数：460 千字
2024 年 7 月第一版 2024 年 7 月第一次印刷
定价：**148. 00** 元
ISBN 978-7-112-25722-5
（36507）

a+u

建筑与都市系列丛书学术委员会
Academic Board Members of Architecture and Urbanism Series

建筑与都市系列丛书
Architecture and Urbanism Series

总策划 Production
国际建筑联盟 IAM　文筑国际 CA-GROUP

出品人 Publisher
马卫东 MA Weidong

总策划人/总监制 Executive Producer
马卫东 MA Weidong

内容担当 Editor in Charge
吴瑞香 WU Ruixiang

助理 Assistants
卢亭羽 LU Tingyu　杨 文 YANG Wen　杨紫薇 YANG Ziwei

特约审校 Proofreaders
熊益群 XIONG Yiqun　果然然 GUO Ranran　寇宗捷 KOU Zongjie

翻译 Translators
英译中 Chinese Translation from English:
樊梦莹 FAN Mengying (pp.16-21, 36-45, 74-85)
林天成 LIN Tiancheng (pp22-35, 46-73, 86-217)
日译中 Chinese Translation from Japanese:
吴瑞香 WU Ruixiang (p15, p225)

书籍设计 Book Design
文筑国际 CA-GROUP

中日邦交正常化50周年纪念项目
The 50th Anniversary of the Normalization of
China-Japan Diplomatic Relations

本系列丛书部分内容选自A+U第578号特辑，原版书名：
ノルウェーとデンマークの建築——自然と向き合うデザイン
Architecture in Norway and Denmark: Design with Nature
著作权归属A+U Publishing Co., Ltd. 2018

A+U Publishing Co., Ltd.
发行人/主编：吉田信之
副主编：横山圭
编辑：服部真吏　Sylvia Chen
海外协助：侯 蕾

Part of this series is selected from the original a+u No. 578, the original title is:
ノルウェーとデンマークの建築——自然と向き合うデザイン
Architecture in Norway and Denmark: Design with Nature
The copyright of this part is owned by A+U Publishing Co., Ltd. 2018

A+U Publishing Co., Ltd.
Publisher / Chief Editor: Nobuyuki Yoshida
Senior Editor: Kei Yokoyama
Editorial Staff: Mari Hattori　Sylvia Chen
Oversea Assistant: HOU Lei

封面图：群岛——G的建筑的模型图，建筑设计：曼蒂·库拉
封底图：诺德霍夫小屋，室内曲面墙壁与天花板，建筑设计：奥斯陆工作室

本书第220页至231页内容由安藤忠雄建筑研究所提供，在此表示特别感谢。

Front cover: Archipelago – Building from G, designed by Manthey Kula. Model photo. Photo courtesy of the Architect.
Back cover: Cabin Norderhov designed by Atelier Oslo. Curved walls and ceilings. Photo courtesy of the Architect.

The contents from pages 220 to 231 of this book were provided by Tadao Ando Architect & Associates. We would like to express our special thanks here.

Preface:

Denmark, Norway and the Nordic School of Architecture

FANG Hai

In all rankings of 20th century modern architecture, the Sydney Opera House always be in the top ten. If asked to name one of the most famous buildings of the 20th century, the Sydney Opera House would be the favorite. It is not only the city landmark in Sydney but also the image card of Australia. However, this famous architectural masterpiece was designed by Danish architect Jorn Utzon[*1], and the chief judge who chose Jorn Utzon's architectural design as the winning proposal of the Sydney Opera House International Design Competition was the Finnish-born American contemporary architect Eero Saarinen[*2]. Therefore, we can see the strong creativity and influence of Nordic architects.

Reviewing the development history of modern architecture and modern design, the Nordic School of Architecture has always shown its strong original spirit and lasting influence. As Mackintosh[*3] from Scotland, Otto Wagner[*4] from Austria, Wright[*5] from the United States, and other pioneering masters began to create modern architecture and design, Finnish Eliel Saarinen kept pace with them and created the Nordic School of Architecture characterized by national romanticism. When modern architects such as Gropius[*6], Corbusier[*7], Mies[*8], and Rietveld[*9] comprehensively promoted the great movement of modern architecture and design, Alto[*10] from Finland was the first to realize the important relationship between modern architecture and regional culture and humanism, which has enabled modern architecture to move towards a healthier and more ecological diversified sustainable developing way. At the same time, it has also made the Nordic School the most important design school in modern design in terms of architecture and design.

Masters of the Nordic School, with unlimited creativity, have come forth in large numbers. For architecture, besides Eero Saarinen and Alto, the Swedish master Asplund[*11], the Danish master Jacobsen[*12], and the Norwegian master Fehn[*13] are all important standard bearers of the Nordic School of Architecture, while Jorn Utzon is regarded as the most outstanding representative of the third generation of modern architectural masters in the epoch-making work "Space, Time and Architecture", which was written by famous architectural historian Sigfried Giedion[*14]. For design, besides Eero Saarinen and Alto; Klint[*15], Wegner[*16], Juhl[*17], and Panton[*18] from Denmark; Tapiovaara[*19], Wirkkala[*20], Kaj Franck[*21], Timo Sarpaneva[*22], Nurmesniemi[*23] from Finland; as well as numerous design masters from Sweden and Norway in various fields of modern design, they have jointly created the most powerful academic school of modern design in the 20th century, and have produced a wide and far-reaching influence on a global scale.

As for the study of architectural theory, the Nordic school also has a large number of masters. For example, Schultz[*24], a Norwegian architectural historian, is regarded as the most influential architectural historian in the West after Giedion for his comprehensive interpretation of the western architectural history and his pioneering research on the spirit of the building site, and the Finnish contemporary architect Pallasmaa[*25] has become one of the most famous architects and architectural critics for his multidisciplinary interpretation of architectural phenomenology and behavioral architecture.

Traditionally, Northern Europe is mainly composed of Denmark, Finland, Norway, and Sweden, and then newly independent Iceland. The four Nordic countries are harmonious but different in modern architecture and design

序言：

丹麦、挪威与北欧建筑学派

方海

在全球各地所有关于20世纪现代建筑的排行中，悉尼歌剧院肯定永远位列前十，如果要选出一座20世纪最著名的建筑物，悉尼歌剧院也是呼声极高的。悉尼歌剧院不仅是悉尼的城市地标，而且是澳大利亚整个国家形象的名片。然而，这个名垂青史的建筑杰作却是由丹麦建筑师约翰·伍重[*1]设计的，而当初力排众议确定伍重的建筑设计竞赛方案为中标方案的悉尼歌剧院国际设计竞赛首席评委，就是祖籍芬兰的美国当代建筑大师艾洛·沙里宁[*2]。由此可见北欧建筑师们旺盛的创造力和影响力。

在现代建筑和现代设计的发展历程中，北欧建筑学派迄今一直表现出强大的原创精神和持久的影响力。当英国苏格兰的麦金托什[*3]、奥地利的瓦格纳[*4]、美国的赖特[*5]等建筑先驱大师开始现代建筑和设计的创造历程时，芬兰的老沙里宁就与他们并驾齐驱，开创了以民族浪漫主义为特征的北欧建筑学派。当格罗皮乌斯[*6]、柯布西耶[*7]、密斯[*8]、里特维德[*9]等现代建筑大师全面推动现代建筑和设计的伟大运动时，芬兰建筑师阿尔托[*10]最早意识到现代建筑与地域文化、人本主义关系的重要性，从而使现代建筑走向更健康、更生态的多元化持续发展轨道。同时也使北欧学派无论在建筑还是设计方面，都成为现代设计最为举足轻重的设计流派。

北欧学派大师辈出，创意无限。在建筑方面，除老沙里宁和阿尔托之外，瑞典大师阿斯普隆[*11]、丹麦大师雅各布森[*12]、挪威大师费恩[*13]都是北欧建筑学派的重要旗手，而伍重更是被著名建筑史家吉迪翁[*14]在其划时代著作《时间、空间与建筑》中誉为第三代现代建筑大师中最为杰出的代表。在设计方面，除老沙里宁和阿尔托之外，丹麦的柯林特[*15]、威格纳[*16]、居尔[*17]、潘东[*18]，芬兰的塔佩瓦拉[*19]、威卡拉[*20]、凯·弗兰克[*21]、萨帕奈瓦[*22]、诺米斯耐米[*23]，以及瑞典、挪威在现代设计诸领域的众多设计大师，他们共同铸就了20世纪现代设计最强大的学术流派，在全球范围内产生了广泛而深远的影响。

在建筑理论研究方面，北欧学派同样大师辈出，挪威建筑史学家舒尔茨[*24]以其对西方建筑史的全面解读和对建筑场所精神的开创性研究，被誉为继吉迪翁之后西方最有影响力的建筑史学家。而芬兰当代建筑大师帕拉斯玛[*25]则以其对建筑现象学和行为建筑学精心而多学科的诠释，成为当代最著名的建筑师兼建筑评论家之一。

传统的北欧主要由丹麦、芬兰、挪威和瑞典构成，后又加入独立不久的冰岛。北欧四国在现代建筑和设计方面，和而不同，各具文化特色，如芬兰旺盛的设计创意，瑞典强悍的科技实力，而丹麦和挪威也同样以其独特的文化创意著称于世。

丹麦在公元8世纪到11世纪的维京时期是一个民风彪悍的“海盗王国”，曾长期侵扰欧洲沿海和不列颠岛屿，也曾建立起庞大的贸易帝国。因此，我们毫不奇怪丹麦现代设计之父柯林特很早就豪迈宣称——全世界最优秀的设计传统都是丹麦现代设计的祖先。丹麦与瑞典一样，亦长期保留着诸多领域领先于全球的科学传统，先有布拉赫·第谷为现代天文学的发展奠定了最翔实的科学基础，后有尼尔斯·玻尔及其创立的哥本哈根学派，对现代物理学尤其是核物理学做出了无可替代的贡献。丹麦的文学艺术同样发达，《安徒生童话》在全球几乎家喻户晓。丹麦的现代设计尤其是家居设计更是长期执全球家具设计之牛耳，他们在现代家具和产品设计的每个方面都创造出非凡的成绩，丹麦设计也成为“美好设计”

because of their cultural characteristics, such as vigorous design creativity in Finland, strong technical strength in Sweden, and unique cultural creativity in Denmark and Norway.

Since ancient times, Denmark has been a pirate kingdom with strong folk customs, which has invaded Britain and France for a long time, and it also established a great trading empire. Therefore, we are not surprised that Klint, the father of Modern Design in Denmark, boldly declared that the world's best design tradition is the ancestor of modern design in Denmark. Like Sweden, Denmark has long kept a scientific tradition leading the world in multiple technosphere. For example, Tycho Brahe laid the most detailed scientific foundation for developing modern astronomy, and then Niels Henrik David Bohr and his Copenhagen School making irreplaceable contributions to modern physics, especially nuclear physics. Danish literature and art are equally well developed, and Andersen's fairy tales are almost household names in the world. Denmark's modern design, especially furniture designs, has long been a global leader in home design. They have made extraordinary achievements in every aspect of modern furniture and product design. Danish design has also become a synonym for good designs, and Danish modern architecture, besides the previous generation of masters such as Jakobson and Utzon mentioned above, has been handed down as well, especially the group of modern Danish architects, who have achieved great achievements all over the world with their outstanding creativity. Some of them are introduced in this book.

From the perspective of national origin, Norway is a close relative of Denmark. It is also a pirate kingdom with a long history, but at the same time, it has established a glorious tradition of wooden constitutive architecture, in which Urnes Stave Church has been listed in the World Cultural Heritage list, so it is easy to understand the extraordinary structural creativity of Norway architects such as Finn from wooden structure to concrete. Norwegians have a natural tradition of exploration. From Fridtjof Nansen to Roald Amundsen, from the North Pole to the South Pole, Norwegian explorers have made indelible contributions to the geographical discoveries of human beings, which endowed Norwegian architects with a natural spirit of exploration in their designs. Norwegian cultural and artistic traditions are equally strong and unique. Playwright Henrik Ibsen is a milestone in modern drama, while great painter Munch has a far-reaching influence. He is not only the standard bearer of modern Painting in Northern Europe but also the ancestor of German Expressionism. These comprehensive cultural traditions have nourished generations of Norwegian architects, and often lead the trend of the times, especially in the construction and spatial form. Therefore, the examples in this book introduce the creative design explorations of contemporary Norwegian architects from several aspects.

Editor's Notes:

[1] **Jorn Utzon**(1918–2008) was a Danish architect, and graduated from the Royal Danish Academy of architectural arts. His most famous design work is the Sydney Opera House.

[2] **Eero Saarinen**(1910–1961) was a Finnish American architect, and graduated from the Architecture Department of Yale University. He is world famous for the Jefferson National Expansion Monument (commonly known as St. Louis Arch), David S. Ingalls Rink and TWA Flight Center, etc.

[3] **Charles Rennie Mackintosh**(1868–1928) was a British Scottish architect, his works belong to the style of Arts and crafts movement, and he is also the main advocate of the British new art movement.

[4] **Otto Wagner**(1841–1918)was an Austrian architect, planner, designer, educator and writer. His style changed from early historicism to budding modernism. He once published *Moderne Architektur*.

[5] **Frank Lloyd Wright**(1867–1959) was one of the greatest archi-

的代名词。而丹麦现代建筑，除了前文已提及的雅各布森、伍重等老一代大师之外，也同样代有传人，尤其是丹麦当代新锐建筑师群体，更是以其卓越的创意在全球各地获得佳绩，本书中所介绍的就是其中的部分代表。

从民族起源而言，挪威是丹麦的近亲，它同样是一个历史悠久的"海盗王国"，但同时也建立起了辉煌的木构建筑传统，其中世纪木构教堂早已列入世界文化遗产，由此，我们可以很容易理解挪威建筑大师如费恩等人由木构到混凝土超凡脱俗的构造创意。挪威人有天生的探险传统，从南森到阿蒙森，从北极到南极，挪威探险家为人类的地理大发现做出了不可磨灭的贡献，也由此使挪威建筑师的设计具有天然的探险精神。挪威的文化艺术传统同样强烈而独特，剧作家易卜生是现代戏剧的里程碑，而画家蒙克更是影响深远，他不仅是北欧现代绘画的旗手，而且还是德国表现主义的鼻祖。这些综合的文化传统滋养着一代又一代挪威建筑师，他们勇于创新，尤其在构造及空间场所形态方面更是时常引领时代潮流。本书中的实例正是从不同侧面介绍当代挪威建筑师的充满创意的设计探索。

编注：

[1] **约翰·伍重**（1918–2008）丹麦建筑师，毕业于丹麦皇家建筑艺术学院，代表作是悉尼歌剧院。

[2] **埃罗·沙里宁**（1910–1961）美籍芬兰建筑师，毕业于耶鲁大学建筑系，代表作是杰斐逊国家扩展纪念碑（俗称圣路易弧形拱门），耶鲁大学英格斯冰球馆和 TWA 飞行中心等。

[3] **查尔斯·雷尼·麦金托什**（1868-1928）英国苏格兰建筑师。他的作品属于工艺美术运动风格，同时他也是英国新艺术运动的主要倡导者。

[4] **奥托·瓦格纳**（1841–1918）奥地利建筑师、规划师、设计师、教育家兼作家。他的风格从早期的历史主义转变为现代主义，曾出版《现代建筑》一书。

[5] **弗兰克·劳埃德·赖特**（1867–1959）美国建筑师，工艺美术运动美国派的主要代表人物。代表作包括建立于宾夕法尼亚州的流水别墅和罗比住宅。他与瓦尔特·格罗皮乌斯、勒·柯布西耶、密斯·凡·德·罗并称四大现代建筑大师。

[6] **瓦尔特·格罗皮乌斯**（1883–1969）德国现代建筑师和建筑教育家，现代主义建筑学派的倡导人和奠基人之一，公立包豪斯学校的创办人。他与弗兰克·劳埃德·赖特、勒·柯布西耶、密斯·凡·德·罗并称四大现代建筑大师。

[7] **勒·柯布西耶**（1887–1965）法国建筑师、城市规划家和作家。他是现代主义建筑的主要倡导者，被称为"现代建筑的旗手"。代表作有朗香教堂、萨伏伊别墅等。他与弗兰克·劳埃德·赖特、瓦尔特·格罗皮乌斯、密斯·凡·德·罗并称四大现代建筑大师。

[8] **路德维希·密斯·凡·德·罗**（1886–1969）德国建筑师。密斯坚持"少即是多"的建筑设计哲学，在处理手法上主张流动空间的新概念。他与弗兰克·劳埃德·赖特、瓦尔特·格罗皮乌斯、**勒·柯布西耶**并称四大现代建筑大师。

[9] **吉瑞特·托马斯·里特维德**（1888–1964）荷兰建筑师、家具设计师。在现代主义设计运动中，里特维德是创造出最多的"革命性"设计构思的设计师。他的代表作有施德罗住宅和红蓝椅。

[10] **阿尔瓦·阿尔托**（1898–1976）芬兰建筑师。现代建筑的重要奠基人之一，人情化建筑理论的倡导者，现代城市规划、工业产品设计的代表人物，同时也是一位家具设计师和艺术家。

[11] **埃里克·古纳尔·阿斯普隆**（1885–1940）瑞典建筑师。在瑞典本土创造了极为优雅且富于诗意的建筑作品。代表作是 1930 年的斯德哥尔摩展览馆和 1940 年的斯德哥尔摩森林火葬场馆。他的作品所蕴含的精神力量给人们无限的想象力。从新古典主义到现代主义，他的作品也代表不同时期瑞典建筑的转变。

[12] **安恩·雅各布森**（1902–1971）丹麦建筑师。同时也是工业产品与室内家具设计师。第一位将现代主义设计观念导入丹麦的建筑师，他将丹麦的传统材料与国际风格相结合，创作了一系列建筑作品。代表作有 Vista 住宅、Rothenborg 别墅、蚂蚁椅、天鹅椅和蛋壳椅等。

[13] **斯维勒·费恩**（1924–2009）挪威建筑师。1949 年，他在奥斯陆建筑学院获得建筑学学位。他曾分别于 1952–1953 年、1953–1954 年在摩洛哥和巴黎学习，并与让·普罗韦合作。在 20 世纪 50 年代，他与阿恩·科尔斯莫还有其他七位年轻建筑师一起成立了奥斯陆挪威进步建筑集团（PAGON 集团）。他的第一个大型项目与盖尔·格伦合作，是 1955 年的尔肯老人之家。1949 年，他在奥斯陆建立了自己的建筑事务所。自 1971 年起，他在奥斯陆建筑与设计学院担任教授，直到 1995 年退休。

[14] **西格弗莱德·吉迪翁**（1888–1968）波西米亚裔瑞士历史学家及建筑评论家。曾任国际建筑会议秘书长。代表作品《空间·时间·建筑》。

[15] **P·V·延森·柯林特**（1853–1930）丹麦建筑师和设计师。被称为"丹麦现代设计之父"。他以设计了哥本哈根格伦特维格教堂而闻名，这是当时丹麦最重要的建筑作品之一。他也以家具设计引领丹麦现代思想潮流。

[16] **汉斯·J·威格纳**（1914–2007）丹麦家具设计师。丹麦最具创意且产量颇丰的一位设计师。他设计的座椅被各大知名设计博物馆收藏，其中包含著名的中国椅。

[17] **芬·居尔**（1912–1980）丹麦家具设计师、建筑师和雕塑家。第一位将丹麦当代风格引入国际的设计师。他的作品被称为"优雅的艺术创造"。

tects in the United States, the main representative of the American School of Arts and crafts movement. His masterpieces include Fallingwater built in Pennsylvania and the world's top Robbie house. Together with Walter Gropius, Le Corbusier and Mies van der Rohe, he is known as the four great masters of modern architecture.

[6] **Walter Gropius**(1883–1969)was a German modern architect and architectural educator, one of the initiators and founders of modernist architectural school, and the founder of the Bauhaus School of Art and Architecture. Together with Frank Lloyd Wright, Le Corbusier and Mies van der Rohe, he is known as the four great masters of modern architecture.

[7] **Le Corbusier**(1887–1965)was a French architect, urban planner and writer in the 20th century. He was the main advocator of modern architecture, known as the "flagman of modern architecture" . His representative works include the Pilgrimage Chapel, the Villa Savoye, etc. He and Frank Lloyd Wright, Walter Gropius and Mies van der Rohe are known as the four great masters of modern architecture.

[8] **Ludwig Mies Van der Rohe**(1886–1969)was a German architect. Mies adhered to the architectural design philosophy of "less is more" and advocates the new concept of flowing space in terms of handling methods. He and Frank Lloyd Wright, Walter Gropius and Le Corbusier are known as the four great masters of modern architecture.

[9] **Gerrit Thomas Rietveld**(1888–1964)was a Dutch architect and furniture designer. In the modernism design movement, Rietveld is the design master who creates the most "revolutionary" design ideas. His representative works include the Schroder House and Red and Blue Chair.

[10] **Alvar Alto**(1898–1976)Finnish modern architect. He's one of the principal modern architecture founders, the pioneer of humanization architecture theory. As the representative of the modern city plan and industrial design, he was also a furniture designer and artist.

[11] **Erik Gunnar Asplund**(1885–1940)was a Swedish architect. He created extremely elegant and poetic architectural works on Sweden mainland. The representative buildings are the Stockholm International Exhibition (1930) and the Woodland Cemetery, Stockholm. The mental strength from his projects brings unlimited imagination and interpretation space to the viewers. His works can represent the Sweden architectural transformation of different periods from the Neoclassicism to the modernism.

[12] **Arne Jacobsen**(1902–1971)was a famous Danish architect in the 20th century and a master of industrial products and interior furniture design. He was the first architect who introduced the modernist design concept into Denmark. He combined Danish traditional materials with the international style and created a series of architectural works. His most famous works include Vista villa, Rothenborg House, ant chair, swan chair, egg chair, and so on.

[13] **Sverre Fehn**(1924–2009)was a Norwegian architect. He received his diploma in Architecture, Oslo School of Architecture, 1949. He had been in Morocco 1952–1953 and Paris 1953 - 1954 for study and working with Jean Prouvé. In 1950s, he had formed the PAGON Group (Progressive Architects Group Oslo Norway) along with Arne Korsmo and seven other young architects. His first major project, in partnership with Geir Grung, is Økern Home for the Elderly in 1955. He established his own architectural practice in Oslo in 1949. He worked as a professor at the Oslo School of Architecture and Design from 1971 until he retired in 1995.

[14] **Sigfried Giedion**(1888–1968)was a bohemian Swiss historian, and architectural critic. He was Secretary-General of Congrès International Architecture Moderne. His representative work is *Space, Time and Architecture*.

[15] **P.V.Jensen Klint**(1853–1930)was a Danish architect and designer, known as the father of Danish Modern Design. He was famous for designing Grundtvig's Church in Copenhagen, generally considered to be one of the most important Danish architectural works of the time.He also led the Danish modern design trend with furniture design.

[16] **Hans J. Wegner**(1914–2007)He was a Danish furniture designer, recognized as the most creative and productive one. The chairs he designed are collected by famous design museums, including the famous "Chinese chair".

[17] **Finn Juhl**(1912–1980)was a furniture designer, architect, and sculptor in Denmark. He was the first famous designer to introduce Danish contemporary design style into the world. His work is called elegant artistic creation

[18] **Verner Panton**(1926–1998)was a Danish furniture designer, architect. The representative work is the Panton chair. He is known as one of the most imaginative and creative design masters of the 20th century. He created a serious of the new form with abstract geometric modeling, which brings furniture and interior design works with futuristic dream space color because of the revolutionary breakthrough and innovation in modern furniture design, research and utilization of new technologies and materials.

[19] **Ilmari Tapiovaara**(1914–1999)was a Finnish furniture designer, one of the first generation leaders of interior architect after World War II. His basic starting point for design and creation is the use of quality rather than self-expression. His representative work is "Domus Chair".

[20] **Tapio Wirkkala**(1915–1985) was one of the most outstanding Finnish modern design masters of the 20th century. He was a landmark who pushed Finnish modern design to the world stage and won international reputation after World War 2. The representative work is the "kantarelli" vase designed in 1946.

[21] **Kaj Frank**(1911–1989)was one of the leading figures of Finnish design after World War II . As a designer and the mentor of several generations of Finnish designers. After graduating as a furniture

designer, he devoted most of his career to the art of ceramics and glass. His representative work is Kilta which advocates rebuilding your dining table.

[22] **Timo Sarpaneva** (1926-2006) was a Finnish designer, sculptor and educator. His pioneering glass products combine art with practical design. Besides glass, he also had deep research on textiles, wood, ceramics and metals. He designed high-grade kitchenware and porcelain, they are both artistic and practical, walked into thousands of households. Sarpaneva's work had enhanced the reputation of Finnish design on the international stage. In the career he had also received a lot of recognition.

[23] **Antti Nurmesniemi**(1927–2003)was a Finnish designer, one of the most important designers of the 20th century, one of the principal figures in developing Finnish design art. His extraordinary contribution includes architecture, furniture, interior, products, plan, photography, and so on. These works are collected by the numbers of the museum. His representative works are the Palace Hotel, Helsinki, the Triennale chair for the Milan Triennale.

[24] **Christi Norberg-Schultz**(1926–2000)was a Norwegian architect, architectural historian, theorist and critic. A leading figure in the research of architecture and urban development in the 20th century. His representative works include *Meaning in Western Architecture, The Concept of Dwelling: On the Way to Figurative Architecture, Architecture: Presence, Language, Place* and so on.

[25] **Juhani Pallasmaa**(1936–)is a Finnish architect, and former professor of architecture and head of the department of Helsinki University of Technology. He held many academic and civic positions, including the director of the Finnish Architecture Museum and the director of the Helsinki Academy of Arts and Crafts from 1978 to 1983. His representative works are *the Eyes of the Skin: Architecture and the Senses, Understanding Architectural*.

[18] **维纳·潘东**（1926–1998）丹麦家具设计师、建筑师。因其对现代家具设计革命性的突破和创新，对新技术、新材料的研究和利用，创造了一系列具有抽象几何造型新形态，带有未来主义梦幻空间色彩的家具和室内设计作品，被誉为 20 世纪最富创造力的设计大师。代表作是潘东椅。

[19] **伊玛里·塔佩瓦拉**（1914–1999）芬兰家具设计师，是“二战”后“室内建筑师”这种新型职业的第一代领袖人物之一。他对设计和创作的基本出发点是使用质量而非自我表现。代表作是多姆斯椅。

[20] **塔比奥·威卡拉**（1915–1985）20 世纪芬兰最杰出的现代设计大师之一，是“二战”后将芬兰现代设计推向世界舞台并赢得国际声誉的代表人物。代表作是 1946 年设计的坎塔瑞丽花瓶。

[21] **凯·弗兰克**（1911–1989）“二战”后芬兰主要的设计大师之一。作为一名设计师和几代芬兰设计师的导师，他以家具设计师的身份毕业，并将职业生涯的大部分时间贡献给了陶瓷和玻璃艺术。代表作是主张“重新建造你的餐桌”的 Kilta。

[22] **蒂莫·萨帕奈瓦**（1926-2006）芬兰设计师、雕塑家及教育家。萨帕奈瓦开创性的玻璃制品将艺术与实用设计融为一体。除了玻璃，他对纺织品、木材、陶瓷和金属也有深入的研究。设计的高档厨具和瓷器兼具艺术性与实用性，走进了千家万户。萨帕奈瓦的作品提升了芬兰设计的国际声誉。

[23] **昂蒂·诺米斯耐米**（1927–2003）芬兰设计师。20 世纪最重要的设计师之一，发展芬兰设计艺术的主要人物之一。他在建筑、家具、室内、产品、平面以及摄影等领域都有卓越贡献，作品被众多博物馆收藏。代表作有赫尔辛基皇宫酒店、米兰椅等。

[24] **克里斯蒂安·诺伯格-舒尔茨**（1926–2000）挪威建筑师、建筑史学家、理论家和评论家。20 世纪世界建筑与城市发展研究的领军人物。代表作《西方建筑的意义》《居住的概念：走向图形建筑》《建筑——存在、语言和场所》等。

[25] **尤哈尼·帕拉斯玛**（1936–）芬兰建筑师。赫尔辛基工业大学的前建筑教授兼系主任。他担任过许多学术和公民职务，包括 1978 年 ~1983 年的芬兰建筑博物馆馆长和赫尔辛基工艺美术学院院长。代表作有《肌肤之目》《认识建筑》等。

FANG Hai

Dean of School of art and design of Guangdong University of Technology, distinguished professor and doctoral supervisor of "Hundred Talent Program" of Guangdong University of Technology.

方海

广东工业大学艺术与设计学院院长，广东工业大学“百人计划”特聘教授、博士生导师。

Architecture in Norway and Denmark: Design with Nature

挪威与丹麦：自然设计

Editor's Words

编者的话

This book, as the title suggests, features Norwegian and Denmark Architects and their works that demonstrate their approach towards designing with nature. Through our conversations with the local architects and writers in this issue, they bring forth a spectrum of sustainability concerns in the region that paints a contrasting picture as compared to the rest of the world. With Norway and Denmark in the forefront facing the forces of nature such as the receding glacier caused by global warming, it encourages them to build architecture that respects and is sensitive to what is around them. And, despite this harshness of nature, their people continue to pride themselves upon their community spirit, creating cities that are inclusive and livable and architecture that is responsive.

(a+u)

在本书中，我们将介绍面向自然进行设计的挪威和丹麦建筑师及其作品。通过与建筑师和撰稿作者的对话，我们意识到在设计的原动力——可持续性上，北欧建筑与世界所形成的鲜明对比，如由全球变暖而导致的海平面上升问题等，这些对他们来说迫在眉睫的自然威胁都存在于他们的日常生活中。对诸如此类的环境问题的考虑和敏感感知，也成了他们进行建筑设计的动机。大自然如此严酷，然而，这样的严酷并不能阻碍他们不断提案创造更宜居的生存空间，表达对社区的关怀和对城市建设的关注。

(a+u)

Essay:
Local Potentials and Global Challenges
Jesper Nygård

论文：
地方潜能与全球挑战
杰斯珀·尼格德

Like so many other cities around the world, Danish cities have undergone dramatic development over the past three decades. Urban populations are both bigger and younger than they were ten years ago. Our largest cities also have the greatest job concentration as well as the highest job growth. In rural districts, by contrast, the population is declining, the average age is increasing, and traditional jobs in agriculture, fishing and manufacturing are dwindling.

These trends contain many challenges and nuances. While populations are rapidly declining in the smallest villages, larger rural towns are growing. And while the increase in population in the largest cities has driven growth and prosperity, neighborhoods have concurrently become less diverse. At the same time, cities of all sizes throughout the country are struggling to mitigate and adapt to the challenges of climate change: heavy rain, rising sea levels and increasingly frequent and more severe flooding. These developments present urgent challenges that call for the rapid development of new and innovative architectural and planning solutions.

Fortunately, this has not created a divide between urban and rural populations. In rural areas, local governments, citizens and business owners are collaborating in a shared effort to adapt to a new reality. The focus is on generating new opportunities by building stronger communities and identity – often through strategic placemaking, building on local strengths and potentials. At the same time, the more densely populated cities and city regions are dedicated to creating sustainable, livable cities with a high quality of life that are capable of adapting to both current and future challenges.

Architecture and planning on every scale, from design to regional planning, have greatly contributed to the Danish welfare society. For generations, our human-centered approach and sensibility to place have shaped our built environment in a way that embodies our democratic and transparent society. Investing in ambitious, iconic architecture whether it is kindergartens, social housing, schools, city halls, public spaces, libraries or sport facilities is part of creating an inclusive society. The built environment ties us together, it shapes our identity, locally and nationally. This approach to developing and transforming our built environment includes a strong focus on placemaking, utilizing the ability of history, culture and local resources. The potentials of a given place will inform new architecture in a transformation of existing buildings and urban spaces, that mitigates challenges and supports growth, identity and quality of life. This is a focus that Realdania, a member-based philanthropic association, promotes through our philanthropic engagement throughout Denmark. A number of the projects I mention in this essay are from my everyday work and supported by Realdania.

Remnants of industrial culture

With a population of about 5.8 million, Denmark has no megacities, and most of the major cities are port cities. In recent decades, the transformation of former industrial port areas into residential, recreational and commercial areas has been an important driver of urban development throughout the country. Former dock areas and industrial buildings hold a unique potential for urban development rooted in the architectural heritage of the busy,

像世界各国的许多城市一样，丹麦的城市在过去30年中也历经了一段戏剧化的发展过程。城市人口较之10年前不断增长并呈年轻化的态势。几座最大的城市集中了最多的工作机会，同时有着最高的就业增长率。相反，在农村地区，人口减少并呈老龄化，农业、渔业和制造业等传统行业逐渐衰颓。

这些趋势蕴含了许多挑战及细微差别。在那些小村庄人口迅速流失的同时，较大的村镇人口在增加。虽然大城市的人口增长推动了其发展和繁荣，但邻里社区的多样性也随之下降。与此同时，全国各地不同规模的城市都在努力缓解和适应气候变化带来的极端天气，如暴雨、海平面上升以及日益频繁加重的洪灾。这些新事态带来了迫切的考验，需要迅速发展出有别于从前的新型建筑和城市规划解决方案。

幸运的是，这并未在城乡人口之间造成鸿沟。在农村地区，地方政府、公民和企业正在共同努力以适应新现实。他们主要通过激活当地优势和潜力的战略性场所营造等方式，建立更强有力的社区和身份认同来创造更多的机会。同时，人口更稠密的城市和区域则致力于建设可持续的、能提供高质量生活的宜居城市，以此来应对当前和未来的挑战。

小到建筑设计，大到区域规划，建筑和城市规划都为丹麦的福利社会做出了巨大贡献。几代人以来，我们基于以人为本的设计准则和对场所的尊重所塑造的人居环境，展现了一个民主而透明的社会。为了构建一个更具包容性的丹麦，我们也投资了一些大胆的标志性建筑，其中既包括幼儿园、社会住房、学校、市政厅、公共场所、图书馆，也包括体育设施。人居环境将人们连接了起来，它塑造了地域和国家的身份认同。在开发和改造人居环境时，我们主要着重于场所营造以及如何充分利用历史、文化与地域资源。一个有潜力的场所会以既存建筑和城市空间的更新为契机而促生新建筑，从而释解一些问题，促进成长，强化身份认同，人们的生活质量也因此得以提升。这也是会员制的公益协会瑞安达尼亚（Realdania）通过在丹麦的公益活动来促进与推广的重点所在。我在本文中提到的许多项目都来自我的日常工作，这些都得到了来自瑞安达尼亚协会的支持。

工业文化遗迹

丹麦总人口约580万，没有特大城市，且几乎所有的主要城市都是港口城市。近几十年来，从旧工业港口区转型为居民区、休闲区和商业区，成了全国城市发展的重要驱动力。对根植于繁忙工业港口建筑遗产的城市发展而言，昔日码头区域和工业港口建筑具有独特的发展潜力。但在新的语境和更广阔的城市生活叙事中，它们提供了具有包容性的公共空间和通往海港的通道。

例如位于奥胡斯的新中心图书馆和城市客厅的多克1号，其建筑由施密特·哈默·拉森建筑师事务所（SHL建筑师事务所）设计，其周边的公共空间则由景观设计师克里斯汀·延森设计。此项目位于历史上的要塞位置，地处河口地带，维京时代的渔民从这里将渔船拖拉上岸。图书馆、景观区和重建的河流构成一个从历史街区到现在，甚至未来的社区纽带。它已成为来自世界各地的学生、家庭和游客的热门旅游地。

同样，在哥本哈根，港口改造也促成了一系列新的集聚场所和空间连接。例如伦德高与特兰伯格建筑师事务所（L&T建筑师事务所）精心设计的丹麦皇家剧院（a+u 09:10）和与其毗邻的奥菲莉娅广场，这片区域已从单调的

industrial port, but in the new context and broader narrative of city life, with inclusive public spaces and access to the harbor front.

One example of this is the new main library and urban meeting place in the city of Aarhus, Dokk1, designed by Schmidt Hammer Lassen Architects, and the public spaces that surround it, designed by landscape architect Kristine Jensen. Located on a historically central site by the river estuary where Viking-age fishermen pulled their boats on land, the library, landscaped areas and resurfaced river form a link from a historical gathering point to present and future communities. It has become a popular destination for students, families and visitors from all over the world.

In Copenhagen too, the transformation of the harbor has resulted in a series of new meeting places and connections. One example is Lundgaard & Tranberg Architects' elegant Royal Danish Playhouse (a+u09:10) and the adjacent square Ofelia Plads, which has been transformed from a drab harbor area to a vibrant cultural venue. A place that Copenhageners not only visit when they are going to the theatre, but where they also come to have lunch in the restaurant, enjoy a cup of coffee on the outdoor patio, which is set on stilts in the water, or take part in the many activities on Ofelia Plads – from Sunday yoga to outdoor concerts.

The former port areas and buildings hold a wide range of exciting potentials. The transformation of industrial structures to contemporary purposes has proven to be one of the most successful ways to create attractive urban environments. And fortunately, in many cities, existing environments and landmark buildings are being repurposed and transformed in ways that connect us with the history and identity of a place, providing a link to the past while embracing the future city.

An example of this is Nordhavn, Copenhagen's North Harbor, one of the biggest and most significant urban development areas in the city, which is currently being transformed from a former industrial port to a new neighborhood. The master plan aims to preserve many of the distinct, historic and architectural qualities of the area. Corn silos, warehouses and workshops have been preserved and repurposed to meet the new residents' daily needs. One example is COBE's conversion of a former corn silo to an attractive residential building with a restaurant on the top floor – a project that was awarded the most prestigious Danish transformation award. Similarly, multiple functions have been combined in the new multistory car park in the area, where the roof combines playground and outdoor exercise facilities with a great view of the harbor front.

In Denmark's smaller cities and towns too, industrial-era sites and buildings hold potential for future urban development. That includes the city of Ebeltoft, where Praksis Arkitekter is currently transforming a malt factory from 1861, formerly the city's main employer, into a new powerhouse for creative professions, service businesses and culture. The transformation from malt factory to culture factory is undertaken with respect for the old industrial building and the cultural environment around it and will be a key driver of urban development in Ebeltoft in the coming years.

The potential of natural features in placemaking

Potentials of place include a wide range of qualities. In cities, it typically includes historical and architectural potentials, while rural settings often have significant cultural and natural qualities that can be reflected in new activities and architecture.

Far from Copenhagen, on Jutland's windswept west coast, the Wadden Sea Centre (See pp. 52–63), designed by architect Dorte Mandrup and landscape architect Marianne Levinsen, is a beautiful transformation that connects local craft traditions with a contemporary expression and a close relationship with the surrounding landscape. The project weaves past and present together in an architectural idiom that is adapted to the rough marshland setting. The Wadden Sea is an exceptional natural site and has been declared a World Heritage site by UNESCO. The architecture of the visitor centre (See pp. 64–71) highlights and promotes the unique qualities, natural environment and cultural history of the area. Thus, it also generates local development, in part in the form of tourism. Staying on the west coast of Jutland, we find another, similar example: the Danish architectural firm BIG's design of the Tirpitz museum near Blåvand. With its spectacular

港口区转型为充满活力的文化场所。哥本哈根人不仅在去剧院时会经过这个地方，他们还会来这里的餐厅吃午餐，在户外的水上露台上喝一杯咖啡，或者参加奥菲莉娅广场上的各种活动，如周日瑜伽或户外音乐会等。

旧时的港口区和建筑充满了发展潜力。将旧工业设施转化成现代功能已被证明是创造有吸引力的城市环境最成功的方法之一。同时幸运的是，许多城市在将现存环境和地标建筑重新设计利用和改造时，都将我们与一个地方的历史和身份认同联系起来，连接过去，也拥抱未来。

其中一个例子就是位于哥本哈根北港的"诺黑文"项目，它是哥本哈根最具规模、最重大的城市发展计划之一，目前正从早先的工业港口转变为新的社区。总规划旨在保留该地区许多独特的、历史性的建筑风貌。玉米筒仓、仓库和车间被保留并重新利用，以满足新的居民的日常需求。例如COBE的改建项目，将玉米筒仓改造成为一栋引人注目的住宅楼，并在其顶层设有一间餐厅，该项目获得了最负盛名的丹麦改造奖。同样地，该地区新建的多层停车场也合并了多种功能，其屋顶结合了游乐场和户外运动设施，并享有海港前的美景。

而在丹麦较小的城镇中，工业时代的遗址和建筑也在未来的城市发展中深具潜力。在埃贝尔托夫特，普拉克西斯建筑事务所目前正着手将该市以前的主要的工业体、一座1861年就有的麦芽工厂改造成由创意产业、服务产业和文化产业构成的新经济体。从麦芽工厂到文化工厂的转变将尊重旧时的工业建筑以及其周围的文化环境。该项目也将成为未来几年埃贝尔托夫特市发展的主要驱动力。

自然环境在场所营造中的潜力

一个区域的发展潜力包含各种各样的品质。在城市中，它通常是历史和留存建筑的潜力；而农村地区通常具有显著的文化和自然条件。这些可以在新的活动和建筑项目中得到体现。

日德兰半岛远离哥本哈根，在其迎风的西海岸，建筑师多特·曼德鲁普和景观设计师玛丽安·莱文森设计的瓦登海中心（第52-63页），将当地手工艺传统以当代的手法来表现，并与周围景观紧紧联系在一起，完成了一次优美的转型。该项目将过去和现在编织成一种适合于沼泽地环境的建筑语言。瓦登海自然条件优越，是联合国教科文组织评定的世界遗产。游客中心建筑（第64-71页）突出并展示了该地区的独特品质、自然环境和文化历史。因此，它也某种程度上以旅游业参与者的身份促进了当地的发展。在日德兰半岛的西海岸，我们发现了另一个类似的例子：位于布拉万德附近、丹麦建筑公司BIG设计的提尔皮茨博物馆。此建筑凭借其独特的外观和适应于微妙沙丘景观的审慎选址，在很短的时间内吸引了成千上万的游客。

在丹麦还分布着其他规模较小的项目，它们都是尊重场地及其特有潜质的出色建筑，并为当地发展做出了贡献。一个很好的例子是位于日德兰岛斯凯恩河沿岸的学习中心，其前身为泵站。斯凯恩河流域于2002年被复原回更为自然的状态，风景优美，游客人数也因此增加。约翰森·斯科夫斯特德建筑师事务所对三个泵站进行了改造和扩建，使游客可以从室内和室外的最佳位置欣赏风景。

上述项目和地点中，景观和自然构成了新的开发项目和建筑的基础，亦如同多特·曼德鲁普在格陵兰岛设计的冰湾中心（第44-51页）。其他项目则试图突出过往历史或建立在当地文化的优势上。由SHL建筑师事务所设计的日德兰半岛北部的文德西塞尔剧院就是基于后者。霍林是一个人口相对稀少的集镇，有着悠久的剧院传统。这个新建筑不仅提供了一个艺术剧院，它的重要性可能更在于通过建立一个吸引当地年轻人随心放松的活力聚会场所，从而提升了城市形象。

无论是文化还是自然环境的潜质，上述建筑和当代丹麦建筑界的共同特点是对文脉的敏锐性和对地方特色，即当地历史、人居环境和身份认同的强烈建筑自觉。

共同创造宜居城市

虽然场地、城市和潜力都是地域性的，但世界各国的城市通常都面临着共同的挑战。

不论现在还是将来，我们都需要为特定场地找到属于它自己的独特解决方案，同时也需要相互学习，并在合理的时候重复使用既有解决方案。这种对经验、思想和解决方案的交流也是瑞安达尼亚协会旨在通过其公益活动努力促进与推广的。因此，我们资助开发了哥本哈根的一座新地标BLOX。它由荷兰建筑事务所OMA设计，建筑本身及其周围的城市空间完善了哥本哈根港中心区域的发展。曾经的工业港口摇身一变，成为人们骑行或在水岸散步的场所。这里的水清澈见底，可以供市民游泳。

BLOX不仅包括丹麦建筑中心，还有网络和创新中心

architecture but discreet location that adapts to the subtle dune landscape, it has already attracted hundreds of thousands of visitors in a very short amount of time.

Denmark is also dotted with projects on a smaller scale, where exceptional architecture with respect for the place and site-specific potentials contributes to local development. A great example of this is the learning centers in former pumping stations along the Skjern River in Jutland. The Skjern River basin was restored to a more natural state in 2002, with an attractive landscape and an increase in visitor numbers as a result. Johansen Skovsted Architects' transformation and extension of three pumping stations lets visitors experience the scenery from indoor and outdoor vantage points.

In the above-mentioned projects and places – as in Dorte Mandrup's Icefjord Centre (See pp. 44–51) in Greenland – landscape and nature form the potential that new development and architecture build on. Other projects seek to highlight a historical past or build on local cultural strong points. The latter is the case for Vendsyssel Theatre in northern Jutland, designed by Schmidt Hammer Lassen Architects. Hjørring is a market town in a relatively sparsely populated area that has a strong theatre tradition. The new building has not only provided a state-of-the art theatre but, perhaps more importantly, strengthened the city by establishing a new vibrant meeting place, where local youth in an informal setting are invited to spend time.

Whether the potential is cultural or natural in character, a common quality of the above-mentioned buildings and much contemporary Danish architecture is a contextual sensitivity and a strong architectural awareness of the potentials of place – of a site's history, built environment and identity.

Co-creating livable cities

Although the sites, cities and potentials are local, the challenges are often global, shared by cities around the world.

Now and in the future, we need to find and develop unique solutions for unique locations, but we also need to learn from each other and reuse solutions whenever it makes sense. This exchange of experience, ideas and solutions is also something that Realdania aims to promote through our philanthropic efforts. That is why we funded BLOX, a new building and destination in Copenhagen, designed by the Dutch firm OMA. The building itself and the urban spaces around it complete the development of the central part of Copenhagen Harbor. A former industrial port where people can now bike or walk along and across the water, which is even clean enough for swimming.

In addition to the Danish Architecture Center, BLOX is also home to the network and innovation hub BLOXHUB, which works to develop solutions for tomorrow's sustainable cities. BLOXHUB is a workplace and network for big and small companies and researchers – from Denmark and all over the world – working with architecture, design and solutions for future cities. The hub serves as a gateway for bringing knowledge into Denmark and dispersing sustainable Danish solutions to the world. We pursue a similar goal with our support for the global urban network C40, where cities around the world – including Tokyo and Yokohama – work together to reduce CO_2 emissions and thus counter climate change.

Many of the global challenges we face call for local solutions in our cities, big and small. Solutions that not only promote the technical transition to a more sustainable society but also a sustainable built environment and inclusive cities with room for homes, workplaces and recreation. Good settings for local communities. Respect for the site and the ability to develop and promote site-specific potentials are important resources in the development of our future cities – from megacities to villages.

The Danish tradition for using architecture to connect people and promote belonging and a shared identity is an agenda that is clearly taken up and realized by contemporary urban developers and architects – to the benefit of current and future city-dwellers. By drawing on the spatial, material, cultural, natural and historical potentials of the site we can create a high level of quality of life in areas and urban environments that have a unique character and a relevant story about past, present and future, which in turn helps to build identity, pride and development.

BLOXHUB的总部。该中心致力于为未来的可持续城市开发提供解决方案。BLOXHUB面向丹麦和世界各地为未来城市提供建筑、设计和企划方案的大小型公司以及研究人员提供工作场所和系统平台。该中心会提供一个将知识带入丹麦并向世界传播丹麦可持续解决方案的通道。我们也在自己所支持的全球城市平台网络C40中追求类似的目标。这一项目旨在通过与包括东京、横滨在内的世界各大城市共同努力，以减少二氧化碳排放为目标，来应对气候变化。

我们面临的许多全球性的挑战都要求在大小城市中寻求当地的解决方案。这些解决方案不仅可以促进社会向更可持续性社会的技术过渡，还可以促进其向可持续人居环境以及具有容纳家庭、工作和娱乐场所的包容性城市发展。无论是特大城市还是乡村，宜人的社区环境、对场地的尊重以及开发、促进特定场地的潜力挖掘是我们未来发展的重要资源。

现代城市开发商和建筑师已经明确地采纳并实现了丹麦一直以来通过建筑建立人们的关系并促进归属感和共同自我认同的传统，这对当前和未来的城市居民皆有裨益。通过利用场所的空间、物质、文化、自然和历史潜能，我们可以在具有独特性并拥有过去、现在和未来相关性的地区和城市环境中创造高水平的生活品质。这些反过来也有助于建立身份认同感、自豪感并促进当地发展。

Jesper Nygård is the CEO of Realdania. He joined the Supervisory Board of Realdania in 2003 and was elected chairman in 2009. Before joining Realdania, he was CEO of KAB, a large Danish non-profit social housing association, from 1996-2013.

杰斯珀·尼格德是瑞安达尼亚协会的首席执行官。他于 2003 年加入瑞安达尼亚董事会，并于 2009 年当选为董事长。在加入瑞安达尼亚协会之前，他从 1996 年至 2013 年担任丹麦大型非营利性社会住房协会 KAB 的首席执行官。

Studio Olafur Eliasson

Fjordenhus

Vejele fjord, Denmark 2018

奥拉维尔·埃利亚松工作室

海湾小筑

瓦埃勒峡湾，丹麦 2018

pp. 22–23: View of Fjordenhus along the harbour. This page: View of building's palatial structure half-submerged in water. Photo by David de Larrea Remiro.

第22-23页：港口沿岸的海湾小筑。本页：半沉于水中、宫殿般的构造。

Diagram／体块生成分析图

Fjordenhus has been an exciting opportunity for us to bring years of research in diverse fields – urban space, light conditions, nature, physical movement, how we use our senses – together in one project that truly melds artistic and architectural vision.

In the design team, we experimented from early on with how to create an organic building that would respond to the ebb and flow of the tides, to the shimmering surface of the fjord, and to the ephemeral qualities of daylight, changing at different times of the day and of the year.

Fundamental to the concept of Fjordenhus is the notion that there is no one ideal position from which to view the building. As you move around and through the structure, your perception of space changes continuously, constantly defined and redefined by your movement. It is the time it takes you to pass into or through the building that defines your experience of space. The level floor grounds you, but at the same time, the walls curve, lean in or out, and perpetuate a sense of movement.

Every line seems negotiable, depending on where you are – your movement makes the building soft; it gives you agency.

Another key element of this project was to create the most immediate relation between a building and its environment. The outer walls, which are normally seen as a membrane between inside and outside, are spaces in Fjordenhus – sometimes the spaces are part of the interior and other times they open to the surroundings as balconies. And there is a porosity to the building: while it stands directly in the water, which permeates parts of the ground floor, the building is shot through with many different openings that frame views of the fjord and the natural elements, which makes the presence of nature felt.

More relational than monumental, Fjordenhus is co-created by its shifting fjord context as well as by the people experiencing it. It makes you conscious of your own presence – conscious not only that you are seeing the building, but that the building is also seeing you.

Text by Olafur Eliasson, artist

“海湾小筑”为我们多年来不同类型的研究，包括城市空间、照明环境、自然、物理运动以及如何运用我们的感官，创造了一个令人激动的机会。这个项目可以称得上真正实现了艺术与建筑的融合。

我们从项目初期便开始思考如何设计一座有机的建筑，来回应潮水的涨落、峡湾中波光粼粼的水面以及全年中随时间不断变化的阳光。

“海湾小筑”设计概念的基础是，我们可以从各个角度欣赏这座建筑，没有最佳角度的区分。当处在建筑周围或穿行于建筑内部时，我们对空间的观感会不断变化，并且这种感受会随着我们的脚步被不断重新定义。这便是定义在建筑中人的空间体验的过程。当驻足水平地面时，墙体会不断卷曲、扭动，持续带给我们空间流动感的体验。

每根线条看起来都是迥异多变的，这取决于我们身在何处。我们的移动将使整座建筑更加柔和生动，而建筑的线条也在引导着我们。

这个项目的另一个重要目标，是建造一座与周边环境可以产生最直接联系的建筑。通常建筑外墙会被处理成分隔内外环境的一层表皮，而“海湾小筑”的外墙却是一个空间，有时这些空间是内装的一部分，有时它们则向周围敞开变为阳台。建筑本身是多孔的，当它矗立于水中，一层的立面上遍布着开口，透过它们人们可以将峡湾及其他自然景观尽收眼底，这样的设计在建筑中保留了自然的存在感。

“海湾小筑”由峡湾的自然环境和人的体验共同实现，它将使你意识到自己的存在，意识到不仅是你在观察建筑，建筑也在观察着你。这样的关联性超越了它的里程碑意义。

艺术家 奥拉维尔·埃利亚松/文

Opposite: Main entrance into the building. All photos on pp. 22–35 by Anders Sune Berg unless otherwise noted.

对页：正面入口。

This page: Light reflecting off from the surface of the water is reflected back onto the ceiling, connecting the building interior with its external surroundings. Opposite: Artworks are integrated into the building itself, creating a formal dialogue with the curvature of the building.

本页：水面反射的光在天花板上再次反射，创造了内部空间与周边环境的联系。对页：艺术作品融入建筑，创造出与建筑的曲线“形态上的对话”。

Fifth floor plan／五层平面图

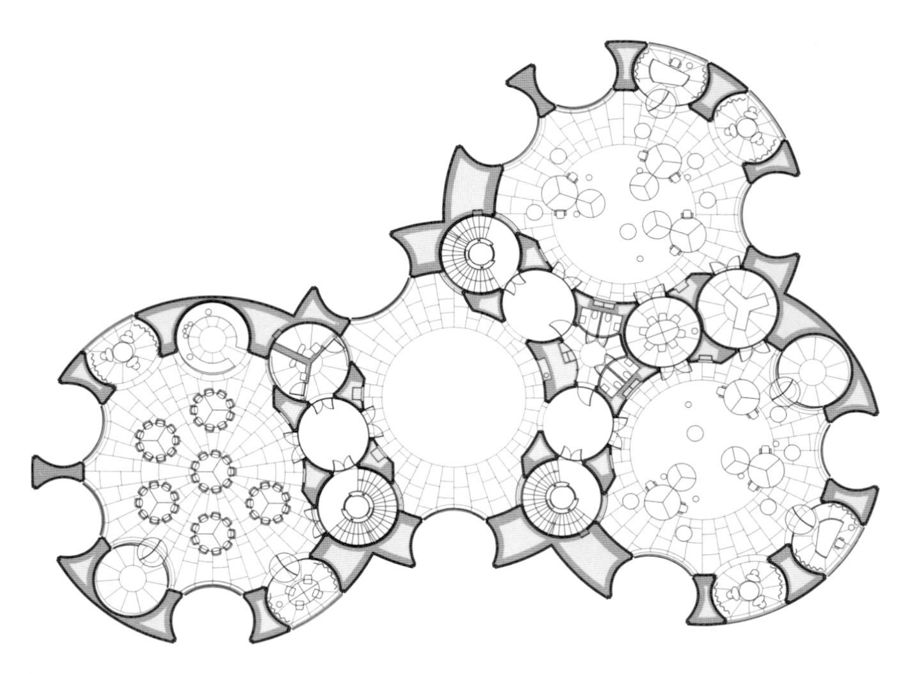

Second floor plan／二层平面图

Ground floor plan (scale: 1/500)／一层平面图（比例：1/500）

Credits and Data
Project title: Fjordenhus
Client: Kirk Kapital
Program: Office
Location: Havneøen 1, 7100 Vejle, Denmark
Completion: 9 June 2018
(Concept Phase: 2009–2011,
Design Phase: 2011–2013,
Building Phase: 2013–2018)
Artist: Olafur Eliasson
Architectural Design: Sebastian Behmann with Studio Olafur Eliasson
Project Architect: Caspar Teichgräber
Local Architect: Lundgaard & Tranberg Architecture
Landscape Architect: Vogt Landscape

Opposite, both images: Floor plans are on different levels, organized around circles and ellipses. This page: Spiral stairs connect to offices on the upper floors.

对页，上图和下图：各层拥有不同的层高，有的平面为圆形，有的为椭圆形。本页：螺旋楼梯通往上层的办公室。

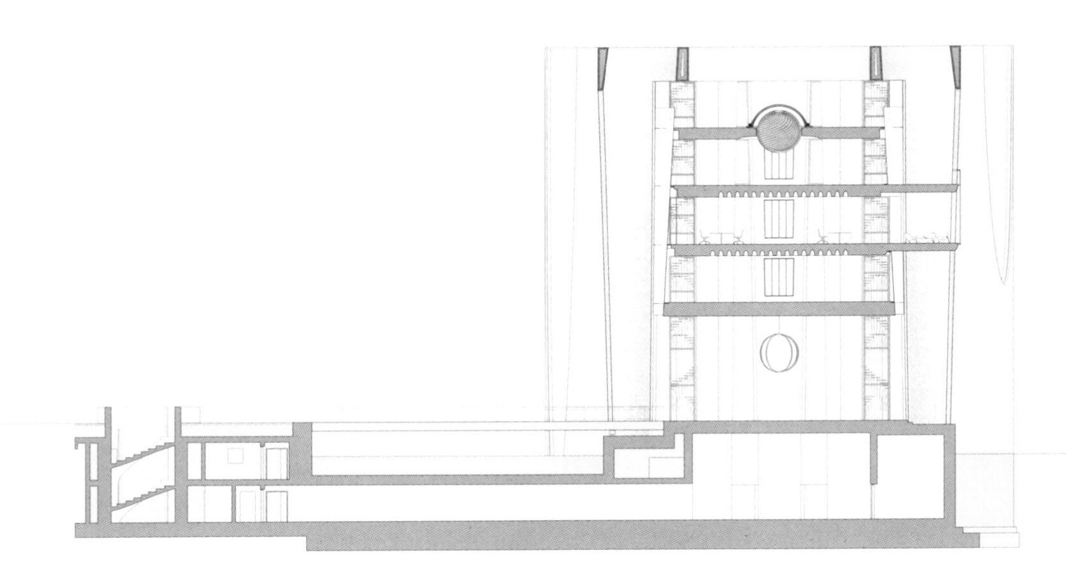

Section (scale: 1/500)／剖面图（比例：1/500）

Opposite: Curved steel framed glass defines the more enclosed spaces of the building while providing visual connections with the surroundings. This page: Overall view of the harbour of Vejle city. Photo by Taylor Dover.

对页：由曲线形钢框固定的玻璃围合出一个封闭空间，同时与周边环境建立起视觉上的联系。本页：瓦埃勒港全景。

Interview: Olafur Eliasson
On Art and Architecture, a Fjordenhus Perspective

采访：奥拉维尔·埃利亚松
艺术与建筑——海湾小筑的视角

a+u: When you received the request from your client to design Fjordenhus, what were the client's needs? Could you share with us what you were expected from your client?

Olafur Eliasson (OE): For a project like this to work, there needs to be a really courageous client because although I had created a lot of artworks and spatial experiments in the past, I had never built a building. I was lucky enough though to have worked for many years with Sebastian Behmann and with the team of architects at my studio. What made Vejle so exceptional is that the Kirk Johansen family, who asked us to build the headquarters for Kirk Kapital, were very interested in creating something that was as much a work of art as a functioning building. During the process of planning and construction, they continually exhibited an incredible respect for artistic vision and for the importance of insisting upon this level of craftsmanship. The results, I believe, are a testimony to their confidence and to the faith they placed in us.

a+u: Between architecture and art, what are your thoughts of their intersection and differences?

OE: Working on Fjordenhus made me realise more clearly than ever before that I am an artist and not an architect, although I enjoy working with architects like Sebastian Behmann, with whom I set up the architectural office Studio Other Spaces. We should not try to turn art into architecture or architecture into art but rather allow them to be two separate things that can be stronger when they work together. I think architects should consult with artists and vice versa to engage in a deeper dialogue about how we build and use space. The fact is, what constitutes architecture today is often far too determined by quantitative considerations and a maximisation of commercial use of space. Art thinks of space in different terms, focusing on qualitative questions such as atmosphere, and what does space do to me, the person moving about that space, and how do I, by occupying and moving through that space, become a kind of co-producer of it. But I wouldn't stop there: the more broadly we expand the conversation, the stronger the results will be. It is necessary when building houses, cities, even countries, to bring in the input of neuroscientists and dancers, poets and parkour athletes, to reconsider how space is constituted, whom it belongs to, and what it can potentially be.

a+u: I believe that Fjordenhus is your first project to make the plan from the beginning to end; include every piece of furniture and brick, you designed in detail by your sketch. When you made a plan and designed these whole things, what kind of image did you have?

OE: In one sense, the building shares so much with the rest of my artistic practice that you can say that it is as much an artwork as everything else I do. It speaks the same language as even the smallest watercolours that I have made, which feature overlapping circles and colour fades. Obviously, a building requires levels of planning and teams of engineers, architects, and construction workers that a watercolour does not. There is also a timescale involved in such a project that means that ideas change and

a+u:当您收到“海湾小筑”的设计委托时，业主的需求是什么？可以与我们分享一下业主对您提出的要求吗？

埃利亚松:尽管过去我创作了很多艺术品，也做过许多空间实验，但我从未建造过房子。这样的项目想要成功，需要有一个很果敢的业主。我很幸运，多年来能与塞巴斯蒂安·贝曼以及我工作室的建筑师团队合作。瓦埃勒之所以如此与众不同，是因为柯克·约翰森家族，也就是委托我们设计柯克资本公司总部的业主，他们对创造兼具艺术性与功能性的建筑非常感兴趣。在规划和建造的过程中，他们一直对我们的艺术视野和坚持高工艺水平的要求极为尊重。我认为最终成果就是他们信任我们的最好证明。

a+u:您如何看待建筑与艺术之间的异同？

埃利亚松:“海湾小筑”的设计使我比以往更清晰地认识到，我是艺术家不是建筑师。尽管我喜欢与塞巴斯蒂安·贝曼这样的建筑师合作，并与他共同建立了他者空间建筑设计事务所(SOS建筑设计事务所)，但我们不应尝试将艺术变成建筑，或者尝试将建筑变成艺术，而应让它们成为两个独立的事物，因为它们一起运作时可以变得更强大。我认为建筑师和艺术家应该相互沟通，就我们如何建造和使用空间进行更深入的对话。事实是，现今建筑的构成往往由数据分析以及空间如何商业利用最大化所决定。艺术以不同的尺度看待空间，着眼于实质上的问题，例如氛围、空间会为在其中工作生活的人们带来什么，以及人们如何在置身其中并度过一段时光的过程中，成为这个空间的共同营造者。但我不会止步于此，我们越扩大对话范畴，成果就会越突出。在建造房屋、城市，甚至国家时，有必要引入神经科学家和舞蹈家、诗人和跑酷运动员的意见来重新思考空间如何构成，属于谁，以及可能成为什么。

a+u:听闻“海湾小筑”是您从头到尾参与设计的第一个建筑项目。从每一件家具到每一片砖瓦，您都用草图进行了详细绘制设计。当您规划并设计所有这些时，您有什么样的预想？

埃利亚松:从某种意义上说，这座建筑与我其余的艺术实践共享了许多类似的思考，可以说它和我所做的其他作品一样，是一件艺术品。它甚至使用了我创作最小水彩画时的相同的语言。那幅画的特色是重叠的圆圈和渐褪的色彩。显然，不同于水彩画，一座建筑的完成需要各方面的计划安排以及工程师、建筑师和工人团队的配合。在这个项目中依然涉及时间尺度，即想法会随时间发生改变与发展，并需要我们进行取舍。但启发该建筑的基本思想和设计原则在我的其他作品也可以看到，包括氛围及环境的重要性，建筑与场地及港口的关系，我们如何通过移动和活动共同创造空间，以及在建筑内部、家具和房间中占主导地位的圆形要素。

a+u:您是如何选择“海湾小筑”的场地的？选择该地点的理由是什么？

埃利亚松:我们在参观了业主方提供的场地并确认了基地选址后，在考虑了要建造的建筑的性质的基础上，提议在水上建造这个建筑。可以这么说，这既是因为柯克兄弟与瓦埃勒峡湾有着紧密的联系且热衷于驾船出海，

This page: View of the Fjordenhus office rooms on the upper floor.
本页：海湾小筑的上层办公空间。

evolve, additions are made, and other things are discarded. But the basic ideas and principles that inspired the building can be found in my other works – the importance of the atmosphere and the relationship to the fjord and harbour where the building is located, considerations of how we co-produce space through movement and activity, as well as the circular forms that dominate in the building, in the furniture and rooms.

a+u: How did you decide the site of the Fjordenhus? Was there any reason or background to choose that place?

OE: We visited the site that was offered to us and after seeing the location and considering the kind of building we wanted to create, we proposed building in the water itself, both because the Kirk brothers feel a close connection to Vejle fjord and are enthusiastic sailors and because we wanted to create a building that is really part of the fjord, is shaped by it, so to speak. Positioning the building in the water creates a situation, where there is a strong connection between the building and the environment that surrounds it. The plaza in front of the building, together with the bridge that connects it to the island, reflects a scale and sense of urban space similar to the situation in a city like Venice, and the grotto-like spaces beneath the building, where the water flows through it, emphasise this porousness between outside and inside, between positive and negative space, between fjord and city.

也是因为我们想创造一座真正属于峡湾并由其塑造的建筑。将建筑安放在水中，会营造出一种情境，建筑与周围环境之间会产生牢固的联结。建筑前方的广场和连接建筑与岛屿的桥梁反映出类似于威尼斯那样的城市空间的尺度和感受。而脚下有水流过的建筑下方的洞窟状空间，强调了内与外、实空间与虚空间、峡湾与城市之间相互融通的通透性。

Olafur Eliasson was born in 1967. He grew up in Iceland and Denmark and studied from 1989 to 1995 at the Royal Danish Academy of Fine Arts. In 1995, he moved to Berlin and founded Studio Olafur Eliasson, which today comprises more than one hundred team members.

奥拉维尔·埃利亚松生于1967年。他在冰岛和丹麦长大，于1989年至1995年在丹麦皇家美术学院学习。1995年，他移居柏林，创立了奥拉维尔·埃利亚松工作室，该工作室迄今已有100多名成员。

Interview: Sebastian Behmann

On Architecture and Art, a Collaboration with Studio Olafur Eliasson

采访：塞巴斯蒂安·贝曼

建筑与艺术——与奥拉维尔·埃利亚松工作室的合作

An Architectural and Artistic Collaboration

Danish-Icelandic artist Olafur Eliasson is a highly prominent figure in contemporary art. Studio Olafur Eliasson, in its current form, was formed around 2000 to produce his works and exhibitions, as well as a series of public commissions, and today employs around 100 people. Its design director, Sebastian Behmann, is an architect by profession, and has worked with Eliasson on projects including the Harpa Reykjavik Concert Hall, the Rainbow Panorama in Aarhus (a+u11:09), Denmark, the Cirkelbroen (Circle Bridge) in Copenhagen, and the Serpentine Pavilion 2007 (in collaboration with Kjetil Thorsen, a+u07:12). In 2014, Eliasson and Behmann founded Studio Open Spaces in Berlin as an architectural counterpart to the art studio. Among their most notable collaborations is Fjordenhus (Fjord House, See pp. 22–35) in Vejle, Denmark, a building not on (as in next to) the harbor, but actually in it, surrounded by water and connected by a pedestrian bridge and an underground passageway. It employs a traditional material with a long history in the region (glazed brick) in a striking contemporary structure with organic curving forms and voids through which visitors can look up at the sky or down at the sea. We spoke with Sebastian Behmann about the project.

His frequent collaborations with Eliasson usually begin with discussions about the project and its environment, not only physical but also social and psychological. For each new project they try to find the right "language," adapting to the site, the client, and the community rather than imposing the vision they have developed thus far on each project in a consistent manner. Fjordenhus features a highly complex shape, which has all the more impact for its minimization of decorative elements. It is the first time they have worked with bricks, for which this region of Denmark is particularly famous, and bricks are relevant to the site – a harbor undergoing large-scale redevelopment – as they have long been used to build massive structures for storing large quantities of goods, and can also withstand the rough environment of the harbor. The adaptation of this time-honored material to a building with a highly contemporary geometry is an example of the process of "translation" typical of the work of Studio Open Spaces.

Linking Inside and Outside, Traditional and Contemporary

In fact, the development of the building's form came before the decision to work with bricks. It emphasized forming transitional spaces between inside and outside, through buffer zones and the voids that make the surroundings visible from the inside. However, the seemingly counterintuitive choice of brick actually reflects a long tradition of complex brick buildings in northern Germany and Denmark, including Neo-Gothic churches found in nearly every city, and other buildings featuring complicated forms but pared-down decoration that highlights the innate qualities of the building material. Today, this tradition goes hand in hand with the latest technology, as many universities and companies in the region employ CNC (Computer Numerical Control) programming in designs for buildings whose bricks are laid by robots.

For Fjordenhus, the studio was concerned with context, with how the building could be part of a

建筑与艺术的合作

来自丹麦和冰岛的艺术家奥拉维尔·埃利亚松是当代艺术的杰出人物。为了创作艺术作品、举办展览以及承接众多公共委托，奥拉维尔·埃利亚松的工作室成立于2000年左右，现已拥有约100名员工。设计总监塞巴斯蒂安·贝曼是一名职业建筑师，与埃利亚松合作过的项目包括雷克雅未克的“哈帕音乐厅”，丹麦奥尔胡斯的“彩虹全景”(a+u 11:09)，哥本哈根的“圆桥”，以及“2007蛇形画廊展馆”(与凯捷·索尔森合作，a+u 07:12)。作为艺术工作室的补充和延伸，2014年埃利亚松和贝曼在柏林成立SOS建筑设计事务所。他们最著名的合作作品之一是位于丹麦瓦埃勒湾的“海湾小筑”(第22-35页)，该建筑“不在港口”，即不毗邻于港口，而是在港口之内被水包围，以人行天桥和地下通道连接。它采用了该地区历史悠久的传统材料(釉面砖)，具有醒目的现代结构与有机曲线形式和空间，游客可以透过它仰望天空或眺望海面。我们就此项目与塞巴斯蒂安·贝曼进行了交谈。

在他与埃利亚松的频繁合作中，讨论通常始于项目及其周边环境，不仅仅涉及物质方面，还包括社会和心理领域。对于每一个新项目，他们都试图找到合适的“语言”以适应现场、客户和社区，而不是将他们迄今产生的固有想法一致强加给每个项目。海湾小筑具有高度复杂的形状，而其极简的室内装饰放大了其复杂形状的影响。这是他们第一次使用砖。在丹麦的这个正在进行大规模重建的海港地区，砖材尤其出名且与地域密切相关，因为长期以来人们都用它建造存储大量货物的大型仓库，它还可以承受港口的恶劣气候环境。将这种历史悠久的材料用于极为现代的几何形态的建筑，是SOS建筑设计事务所“转译”的典型示例。

连接内与外、传统与现代

实际上，建筑形态在决定使用砖材之前就已经设计好了。它强调通过缓冲区和从内部可见的周围环境的空隙，在内外之间形成过渡空间。然而，看似反直觉的砖材实际上反映了德国北部和丹麦砖混建筑的悠久传统，例如几乎每个城市都能找到的新哥特式教堂，以及其他拥有复杂形式和精简装饰且突出材料本质的建筑。如今，随着该地区的许多大学和公司在建筑设计中采用计算机数控(CNC)编程并通过机器人铺设砖块，这一传统如今被最新技术传承。

对于“海湾小筑”，工作室关注的是场地文脉，即该建筑作为该城市的半个港口重新开发的核心，如何成为瓦埃勒不断变化的港口中新叙事的一部分。同时，另一半港口也有着类似的未来计划。贝曼说，“形成一个想法，一个陈述，一个隐喻……对从港口城市到与水相关的现代工业城市的转变十分重要。”另一个重要因素是建筑本身与水的关系，建筑没有建在水边，而是给居住者和访客(私人拥有的建筑底楼向公众开放)如同在船上感受海浪和风的体验。

赫赫有名的业主

奥拉维尔·埃利亚松艺术家的身份为人熟知，但他却越来越多地参与到建筑创作中。可以看到，他的建筑项目对两个领域均有涉及，是在艺术家和建筑师之间“非常富有成效的往复过程”的另一成果(贝曼语)。就像任何委托一样，业主的动机也至关重要，这次的业主是丹麦最具代表性的公司之一，也是在日本和世界各地都广为人知的公司：乐高集团。乐高积木玩具的发明者和公司创始人奥尔·科克·克里斯第森先生的孙辈三兄弟共同领导着乐高基金会，他们的宗旨是“启迪和培养未来的建造者”。当家庭基金会正在寻求新建一座建筑，并得知瓦埃勒正在出售位于瓦埃勒

new narrative for Vejle's changing harbor as a centerpiece of the city's redevelopment of half of the harbor, with the other half also slated for similar treatment in the future. According to Behmann, "it was important to form an idea, a statement, a metaphor... for the narrative of the city, as a transformation from a harbor city into a kind of contemporary industrial city that relates to the water." The other important element was how the building itself relates to the water, not standing beside it, but giving occupants and visitors (the ground floor of the privately owned structure is open to the public) an experience like being on a boat, hearing crashing waves and wind.

High-Profile Clients

Olafur Eliasson is widely known as an artist but has become increasingly involved with architecture, and the building can be seen as having one foot in each of the two disciplines, another outcome of "a very fruitful ping-pong process of back and forth" (Behmann) between artist and architect. As with any commission, the motives of the clients were of crucial importance as well, and in this case the clients are scions of one of Denmark's most iconic companies, instantly familiar in Japan and around the world: The LEGO Group. Three brothers, grandsons of the LEGO® construction set inventor and founder Ole Kirk Christensen, head the LEGO Foundation, which aims "to inspire and develop the builders of tomorrow." The family foundation was seeking a new building, learned that the city of Vejle was offering this site on Vejle Fjord for sale, and decided to commission a building here that would make a statement, give something back to the world, and convey something of why the brothers choose to make their home in Vejle and the values they share as a family. After looking at various Danish and Scandinavian architectural firms, they remained unable to find one they could sufficiently trust to honor those values, and looked a bit further abroad to Berlin-based Studio Other Spaces, whose projects in Denmark (the Copenhagen bridge, the Rainbow Panorama) and Reykjavik concert house are well known. It was not a great leap from the LEGO Foundation to compatriot Eliasson, with his track record of involvement with architecture in Denmark.

Crossing the Boundaries of Disciplines

Why, some might ask, does an artist get involved in architecture? What sustains the collaboration between Behmann and Eliasson? They share an interest in space and advanced geometry, Eliasson's grasp of which Behmann calls "quite exceptional for an artist that has no architectural training." Behmann recalls that in the early years of their collaborations, which go back to 2001, Eliasson was often traveling and they frequently had discussions of highly complex geometries on the phone, discussions that showed the artist's extraordinary capacity for structural thinking. Meanwhile, architects are often merely commissioned to design a certain kind of structure at a certain site and are not responsible, or particularly concerned with, its aesthetics and how people will experience it. Behmann cites "holistic thinking" as architects' greatest strength, and says they need both to regain freedom to create entire spaces, not simply structures, and to take part in solving the urgent

峡湾的这处土地时，他们决定在此建造一座建筑，并希望通过该建筑的发声来回馈世界，并传达三兄弟在瓦埃勒安家的缘由以及他们作为一个家族共同拥有的价值观。在考察了丹麦和斯堪的纳维亚的各家建筑公司之后，他们仍然找不到合适的公司来实现这些价值观。于是他们把目光投向了柏林的SOS建筑设计事务所，他们在丹麦的项目（哥本哈根桥、彩虹全景）以及在雷克雅未克的音乐厅都很有名。考虑到丹麦同胞埃利亚松在丹麦建筑领域的业绩，乐高基金会委托他们也就不足为奇了。

跨界

有人可能会问，为什么艺术家会参与建筑设计?是什么维持了贝曼和埃利亚松之间的合作?那是因为他们对空间和高级几何学有着共同的兴趣。贝曼评价埃利亚松的悟性，“对于没有经过建筑训练的艺术家来说非常出色”。贝曼回忆道，早在2001年他们的合作初期，埃利亚松经常出差，他们经常在电话上讨论高度复杂的几何形状，这些讨论表明这位艺术家具有非凡的结构思维能力。建筑师通常仅被委托在特定地点设计某种构筑物，对结构的美感以及人的体验不承担责任，或者说，不以为意。贝曼认为“整体思维”是建筑师最大的强项。他说，他们既需要重新获得创造整个空间（而不仅仅是结构）的自由，又需要参与解决未来的紧迫问题。建筑并不凭空存在，而始终与其他行业紧密相连。人们倾向于对“涉足”彼此职业的艺术家或建筑师有所怀疑，但海湾小筑是该团队消除这种怀疑的一个尝试。

有所作为

他们在埃塞俄比亚的亚的斯亚贝巴，一个远离北欧的地方，完成了一个大型项目（梅莱斯·扎纳维纪念公园）。在这个项目中，他们试图在一个新发展的、全球化推进（特别是受中国影响很大）的国家中融入非洲特色。亚的斯亚贝巴是一个快速发展的城市，有大量外国资本注入，但新的建设忽略了传统埃塞俄比亚建筑的价值。此项目海拔2,300米到2,500米，拥有可以将植被、树木或公园整合到建筑中的理想气候，类似于巴西，但蚊虫较少。根据传统，像门廊这样可用来招待客人的空间一直是连接内部和外部之间重要的中间区域。相比之下，新的建筑倾向于用玻璃幕墙来隔绝建筑，这就是SOS建筑设计事务所在这个项目中试图改变之处。他们采用了当代流行的混凝土材料，同时避免了欧洲中心主义的建筑手法。

在其他领域，埃利亚松通过他的公司“小太阳”涉足非洲和其他发展中地区，该公司生产的太阳能灯可供那些居住在缺少电网地区的人们使用。电力并不是在任何地方可以随意使用的，它在某些地方更易得，这其实反映了全球的权力结构。但基于太阳能的分布式能源系统，让学童们能够在晚上做作业，也让他们了解未来是掌握在自己手中的。小太阳公司的命名也是源于这一愿景。他们的工作也引起了非洲民间组织、政府领导人和联合国的关注。作为他在该地区工作的扩展，埃利亚松还为埃塞俄比亚带来了为期半年的欧洲艺术课程。

开始的地方：20世纪90年代的柏林

塞巴斯蒂安·贝曼既是奥拉维尔·埃利亚松工作室的设计总监，也是埃利亚松另一个公司SOS建筑设计事务所中的合作伙伴。两者截然不同，但相互联系。丹麦是一个无论在建筑、家具或其他领域的设计中，对质量都要求极高的国家，并且对建筑有很强的社会民主态度。这使人们强烈地意识到视觉吸引力和审美经验在建筑中的重要性，而不一定与

problems of the future. Architecture does not exist in a vacuum but is always profoundly connected to other professions. There is a tendency to look askance at artists or architects who seem to "dabble" in one another's professions, but Fjordenhus is one example of the team's attempts to put such skepticism to rest.

Making a Difference

Far from northern Europe, in Ethiopia, they completed a major project in Addis Ababa (Meles Zanawi Memorial Park), in which they sought to incorporate African identity in a country where new development has a markedly globalized, and specifically Chinese influence. Addis Ababa is a fast-growing city with a major influx of capital from foreign investors, but new construction ignores the values of classical Ethiopian architecture. At an elevation of 2,300 or 2,500 meters, it has an ideal climate for integrating plants, trees or parks into architecture, similar to Brazil but without many flies or mosquitoes. And traditionally spaces like porches have been important intermediate zones between inside and outside, where guests were entertained. By contrast new architecture tends to shut a building off with a glass curtain wall, and this was what Studio Open Spaces sought to counteract with their project, which employed the contemporary and prevalent material of concrete while avoiding a Eurocentric approach to architecture.

Eliasson is involved in Africa and other areas of the developing world in another way, through his company Little Sun, which produces solar lamps usable by those without access to the power grid. Electricity is not something flowing freely everywhere; its availability in some places more than others reflects global power structures, and a decentralized energy system based on solar energy, while enabling schoolchildren to do their homework in the evenings, also empowers them to understand that the future is in their own hands. This vision went into the naming the company Little Sun, and its activities have drawn the attention of African NGOs, government leaders, and the UN. As an extension of his involvement with the region, Eliasson brought a European art class to Ethiopia for half a year.

Where it All Began: Berlin in the 1990s

Sebastian Behmann is both design director of Studio Olafur Eliasson and partner with Eliasson in Studio Other Spaces, a separate company. The two are distinct but linked. Denmark is a country where design quality matters, in architecture as well as furniture and other areas of design, and there is a strong socially democratic approach to architecture, leading to a strong sense of the importance of visual appeal and aesthetic experience in architecture, if not necessarily connected to art per se. The nation is a world leader in design of public space and is also at the forefront of cutting-edge technology, which helps drive its design.

Collaboration between artist and architect began not in Denmark but in Berlin, where both were living in the 1990s. Post-Wall Berlin was full of empty spaces and disused lofts, making it ideal for artists seeking studios and exhibition venues, and turning it into an art-world capital. At the same time, a surplus of buildings meant the environment was as tough for architects as it was promising for artists, and Behmann recalls that he was forced to venture into other professions to survive, collaborating with artists to help them produce works and models, and even distancing himself from his architectural practice for a time. This led to the collaboration with Eliasson that continues to bear fruit to this day, including Fjordenhus, a new landmark on the Danish shore that puts Olafur Eliasson and his studio more firmly on the architectural map.

艺术本身有关。丹麦在公共空间设计领域领先全球，同时也站在尖端技术的最前沿，这有助于推动其设计的发展。

艺术家与建筑师之间的合作并非始于丹麦，而是始于20世纪90年代的柏林。柏林墙倒塌后到处都是空地和废楼，这非常适合寻求工作室和展览场地的艺术家。他们将这里转变为艺术之都。同时，过剩的建筑意味着一个对艺术家来说充满希望，但对建筑师来说举步维艰的环境。贝曼回忆说，他被迫冒险尝试其他职业以求生存，与艺术家合作、帮助他们制作作品和模型，甚至脱离了自己的建筑实践一段时间。这些经历促成了与埃利亚松持续至今的长久合作，成果包括“海湾小筑”这座丹麦海岸上的新地标，这巩固了奥拉维尔·埃利亚松和他的工作室在建筑界的地位。

Sebastian Behmann, born in Germany in 1969, has worked with Olafur Eliasson since 2001 and is head of the department of design at Studio Olafur Eliasson, as well as co-founder of Studio Other Spaces. Major projects with Eliasson include the Serpentine Gallery Pavilion 2007 in London, Cirkelbroen (The circle bridge) in Copenhagen (2015), and Fjordenhus in Vejle, Denmark (2009–18), in addition to numerous installations, pavilions, and international exhibitions.

塞巴斯蒂安·贝曼于1969年出生于德国，自2001年以来一直与奥拉维尔·埃利亚松合作，是奥拉维尔·埃利亚松工作室设计总监，也是SOS建筑设计事务所的共同创始人。与埃利亚松合作的主要项目包括伦敦的2007蛇形画廊展馆，哥本哈根的“圆桥”（2015年）和丹麦瓦埃勒的“海湾小筑”（2009-2018年），以及众多装置、展馆和国际展览。

Dorte Mandrup

Icefjord Centre + Venice Biennale 2018

Ilulissat, Greenland, Denmark 2016–

多特·曼德鲁普

冰湾中心+2018威尼斯双年展

丹麦，格陵兰岛，伊卢利萨特 2016-

On the western coast of Greenland lies the massive Sermeq Kujalleq glacier.

For more than 250 years, glaciologists have studied the ancient glacier and its daily production of immense amounts of ice, and it remains an ideal spot for scientific observation of climate changes.

Dorte Mandrup has designed the new Icefiord Centre in Ilulissat to blend in with the impressive landscape, while offering local residents, tourists, and climate researchers the ultimate vantage point from which to absorb the historic atmosphere of the Icefiord.

The Icefiord Centre will tell a story of ice, of human history and evolution on both a local and global scale.

The Icefiord area carries 4,000 years of cultural heritage and is essential for today's understanding of climate changes.

Credits and Data

Project title: Icefjord Centre + Venice Biennale 2018
Client: Government of Greenland, Avannaata Kommunia and Realdania
Program: Visiting center, learning and exhibition space
Location: Ilulissat, Greenland, Denmark
Architect: Dorte Mandrup
Landscape Architect: Arkitekt Kristine Jensens Tegnestue
Area: 900 m^2

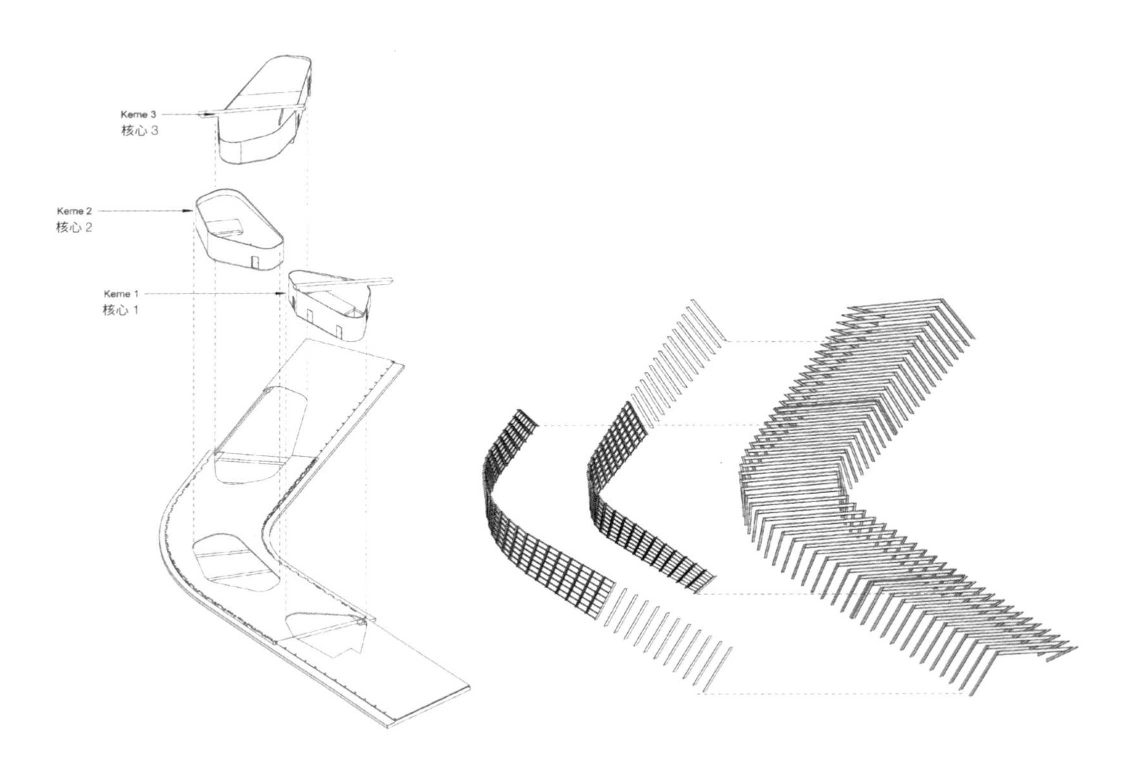

Exploded axonometric drawing／分解轴测图

pp. 46–47, and p. 48: Installation for the exhibition. It focuses on enhancing the understanding of the unique natural conditions that defined the building – more than a mere presentation of the building. p. 49: Entrance before entering the exhibit with the installation. All photos on pp. 46–49 by Adam Mørk unless otherwise noted. Opposite: Render view of the Icefjord centre against the landscape. This page, both images: Render view of the Icefjord Centre. The roof forms an aerodynamic shape to prevent snow build. p. 52: Interior render of the exhibition space. p. 53: Interior render of the viewing gallery.

第46-47页和第48页：展览模型。它不仅仅是为了展示建筑的外形，也在告诉人们，这座建筑是凭借对自然条件的高度理解而诞生的。第49页：进入展厅入口前的艺术空间。对页：冰湾中心与周边环境的效果图。本页，上图和下图：冰湾中心的效果图，屋顶为了防止积雪，采用了空气动力学造型。第52页：展厅的室内效果图。第53页：观光长廊的室内效果图。

在格陵兰岛西岸，分布着巨大的雅各布港冰川。

过去250多年以来，冰川学家们对古代冰川以及它每日生成的巨大冰量做了大量研究，现在这里变成了一个科学观测气候变化的理想据点。

多特·曼德鲁普在伊卢利萨特设计了一座“新冰湾中心”，建筑与当地令人印象深刻的地形融为一体，同时也为当地居民、旅行者和气象学家们提供了一个可感受和融入冰湾地区历史氛围的绝佳场所。

“冰湾中心”讲述了一个故事，一个关于冰的故事，一个关于人类历史的故事，也是一个同时发生在当地和全世界的变革的故事。

冰湾地区拥有的4,000年历史文化遗产，对于帮助我们了解当代气候变化是不可或缺的宝贵财富。

Plan (scale: 1/600)／平面图（比例：1/600）

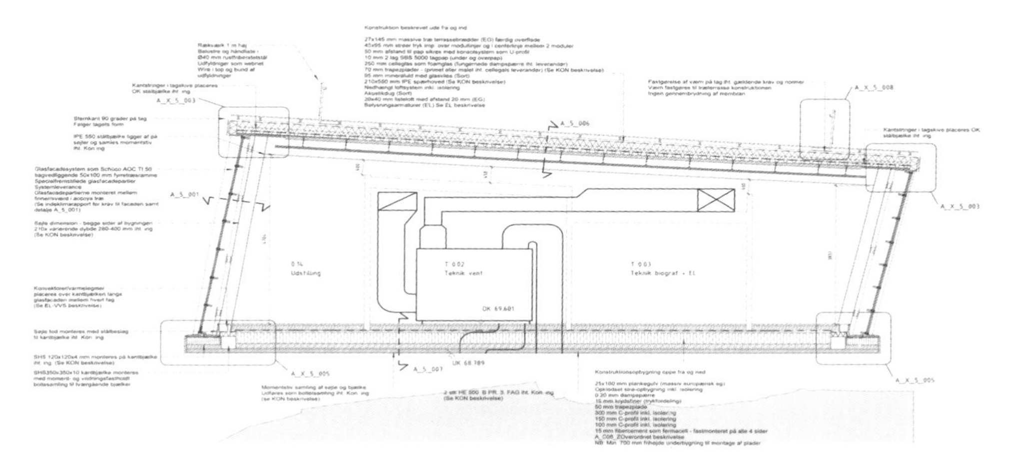

Section EE／EE剖面图

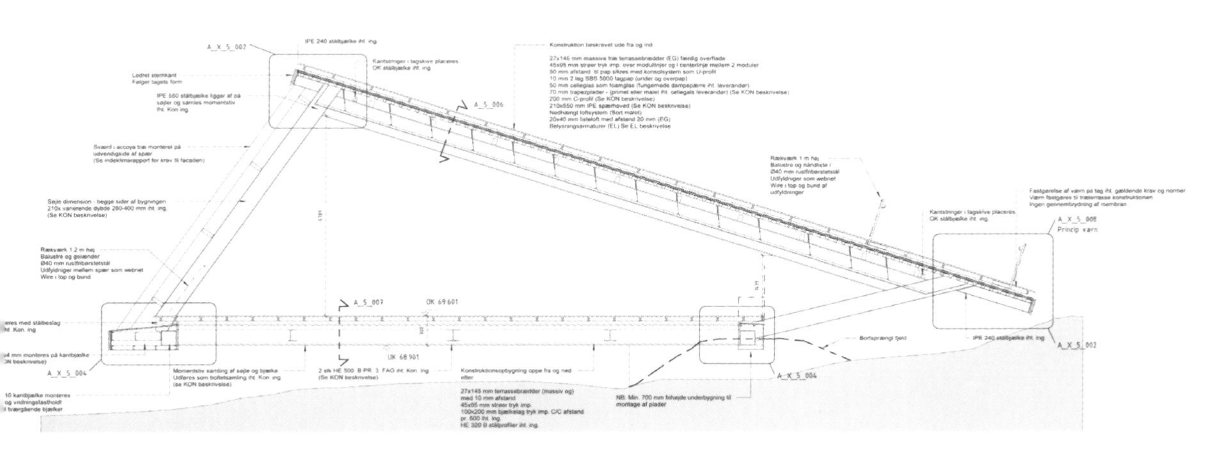

Section FF (scale: 1/350)／FF剖面图（比例：1/350）

Dorte Mandrup

Wadden Sea Centre

Ribe, Denmark 2017

多特·曼德鲁普

瓦登海洋中心

丹麦，里伯 2017

pp. 54–55: View from the entrance looking towards the horizon. This page and opposite: Front view of the building. Its soft, long and clear profile contrast against the Wadden Sea's infinite horizon. All photos on pp. 54–65by Adam Mørk.

第54-55页：从入口看向地平线。本页及对页：建筑的正面，柔和、清晰且修长的轮廓与瓦登海无尽的海岸线形成了对比。

The unique Wadden Sea intertidal zone is a joint venture between human enterprise and the dynamic forces of nature. Since the Iron Age, humans have chosen to settle here at the edge of the generous marsh lands on human-made hill islands – created for protection against the tide. Viking farms and villages, built out of wood and thatched with straw, were opportunely placed here, close to Ribe, the first trade city in Scandinavia.

The Wadden Sea Centre is an extension to an existing, angled exhibition building from 1995. To ensure a cohesive layout and an easy flow, the extension has been designed to partly embrace the existing building in order to create a larger sense of unity. The four-winged farm typology refers to the original buildings of the region.

Traditional and regional craftsmanship

The use of water reeds as cladding for roofs has been in use since the Viking Age – and the craftsmanship remains essentially the same. From a distance, the exhibition centre has the appearance of a large thatched farm emerging as an island in the landscape. The soft, textural reeds are sculpturally processed with long precision cuts in order to create eaves, covered areas, and intersections between diagonal and vertical surfaces.

The entrance is placed in continuation of the existing building's gable. This ensures a circular flow in the exhibition, easy connection to the existing café, and a meeting and service area with the least possible amount of conversion. A sense of arrival is conveyed with a diagonally positioned ramp leading to a covered terrace that runs the full length of the building.

Conjoining exhibition and architecture

The exhibition depicts life in and around the Wadden Sea including the cycle of migratory birds through seven different rooms and themes. It is shaped by various spatial geometries and moods. Different types of daylight and connection to the outdoor spaces is an important part of the experience. The exhibition space has been created in collaboration with the exhibition architect to create the optimal interplay between exhibit and architecture.

The teaching area has two large classrooms and outdoor storing spaces for equipment. It is placed so that teaching and field work have sufficient space and access to the outdoor areas. Two covered areas provide shelter for students and large groups to convene and eat their lunch.

The existing brick building is clad with wooden slats on both the roof and the facades, and new and differently proportioned oak windows have been added to enhance the building's visual coherence.

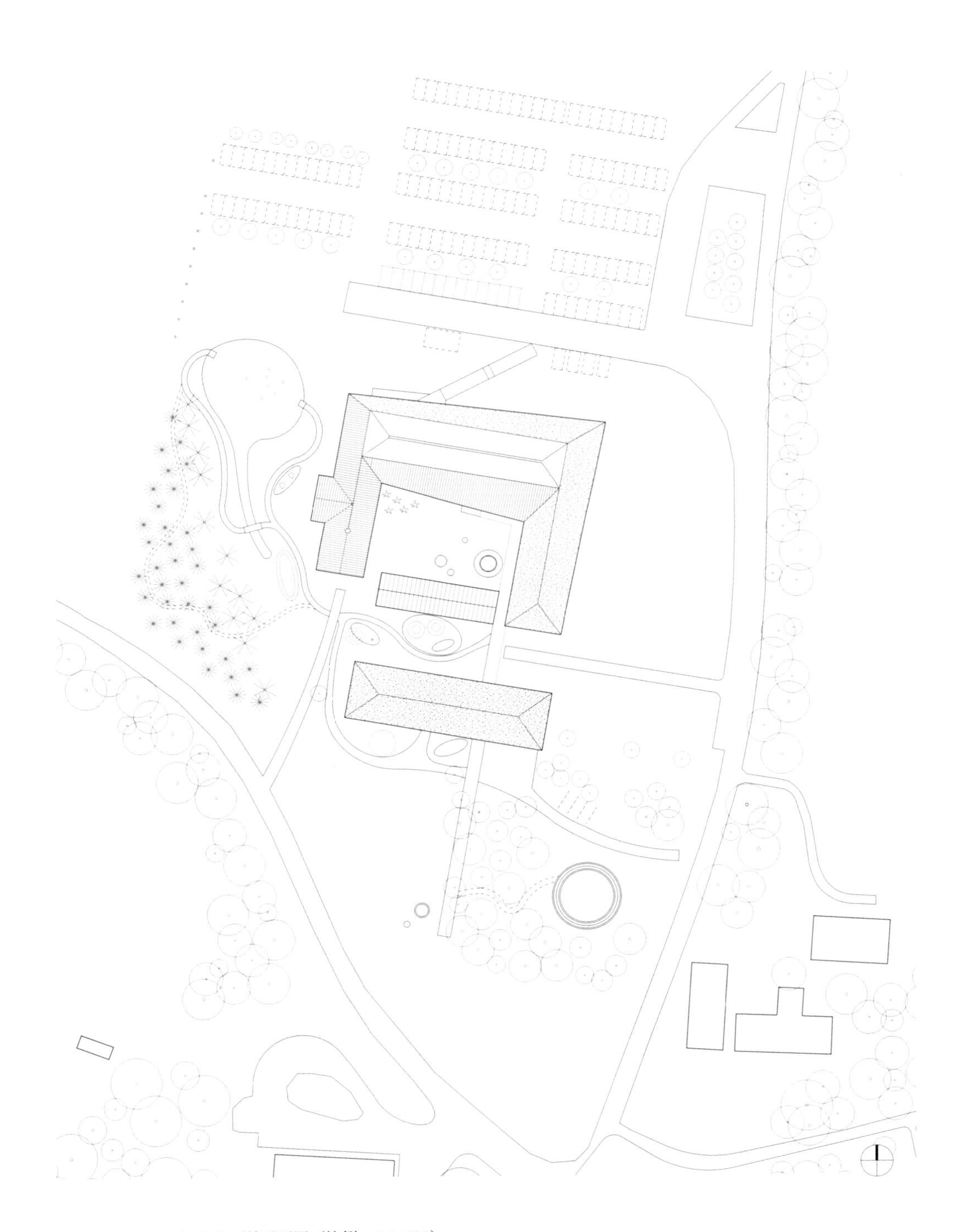

Site plan (scale: 1/1,500) ／总平面图（比例：1/1,500）

这片属于瓦登海的独特潮汐带，是人类的冒险心与大自然不断变化的力量共同作用下的产物。从铁器时代以来，人们便在此堆建起丘陵状的小岛，并在岛上定居下来，目的是保护肥沃的湿地免受潮汐的侵扰。于是，维京人的农场和用木材、芦草建成的村庄便在斯堪的纳维亚第一座贸易城市——里伯的附近扩散开来。

“瓦登海洋中心”是一座建于1995年的既有建筑的增建。它拥有倾斜的屋顶造型。为了让布局更加紧密、流线更加简单，增建部分被设计成部分环保原建筑的形式，以产生更强烈的统一感。四面围合的农场类型则是参考了本地传统的建筑形式。

传统地方工艺

利用芦草覆盖屋顶表面的工艺可以追溯到维京时代，而这门工艺几乎被完好地保留至今。从远处看，展览中心的外形如巨大的草屋农场，像一座岛屿般浮现在周围的景色之中。原本质地柔软的芦草经过雕刻般的加工以及精确的切割，用以制作屋檐、遮盖区以及倾斜面与垂直面的接口。

入口被设置在原建筑山墙的延伸处。这样的设计可以更好地确保展览的环形流线，并在做最少改建的情况下方便人们前往现有的咖啡厅、活动区以及服务区。一条贯穿建筑全长、斜向铺设的通道与一个带顶的平台相连，给游客一种到达终点的暗示。

展览与建筑的结合

这里的展览陈设于7个不同主题的展厅之中，展示了包括候鸟迁徙在内的瓦登海域内外的各种生命活动。展览内容通过多种空间形状和氛围来呈现，其中一个重要的部分是不同自然光和与室外空间的连接。为了实现建筑和展展品之间相互配合的最佳效果，这些展示空间是与专门从事展览设计的建筑师们共同设计完成的。

教学区有两间大教室和一个用于堆放设备的室外储存空间，这样的布局确保了教学和实践活动拥有充足的空间，同时还有一条通往户外区域的通道。两块带屋顶的区域是学生和多人团体聚会的餐饮场所。

原有砖砌建筑的屋顶以及立面被木板覆盖，改建过程中加入了不同大小的橡木窗，这增强了整座建筑的视觉统一性。

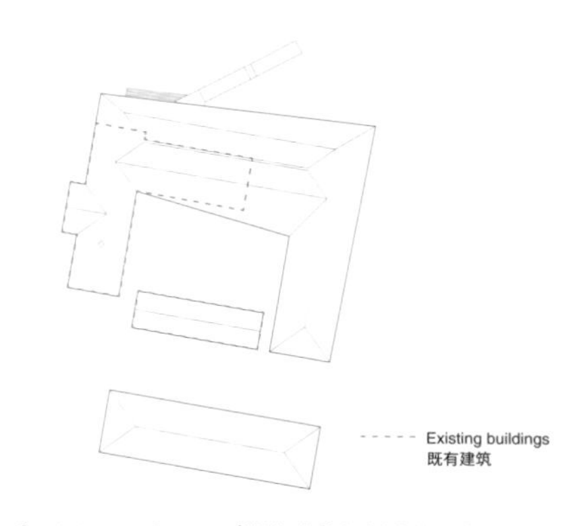

Diagram of existing and now ／既有建筑与新建筑示意图

Credits and Data
Project title: Wadden Sea Centre
Client: Municipality of Esbjerg
Program: Exhibition, Café, Education and Communication
Location: Okholmvej 5,6760 Ribe, Denmark
Completion: 2017
Architect: Dorte Mandrup
Engineer: Steensen & Varming and Anders Christensen
Landscape Architect: Marianne Levinsen Landskab
Area: 2,800 m²

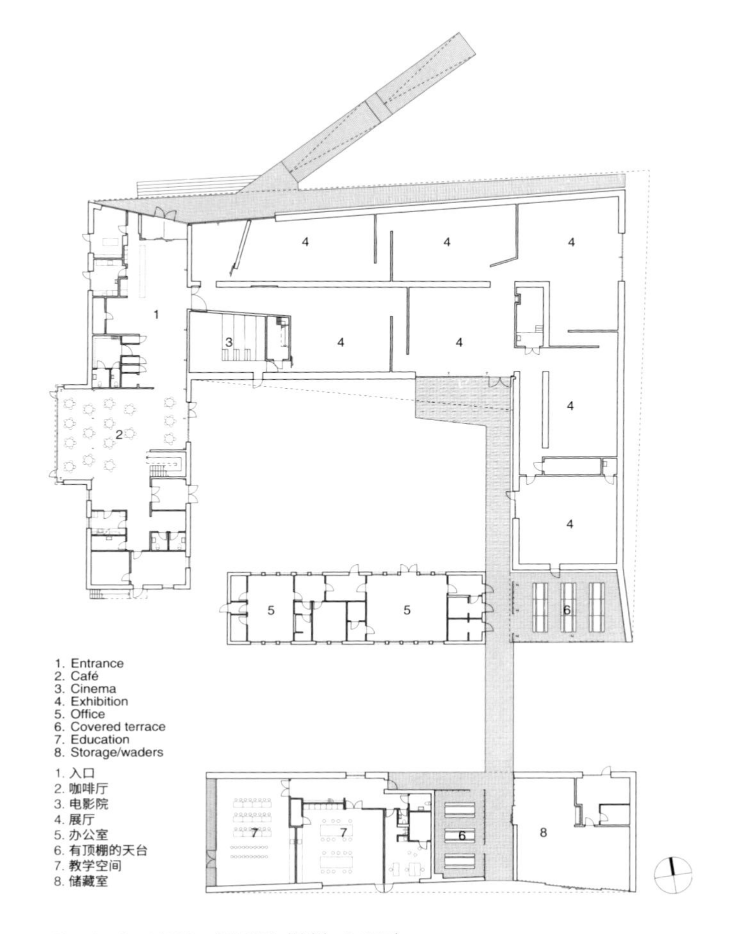

Plan (scale: 1/800) ／平面图（比例：1/800）

p.58: View of the inner landscape garden. Opposite: View of the Wadden Sea landscape.

第58页：内部的景观花园。对页：瓦登海的景色。

TRÆK

Opposite: Exterior view looking into the exhibition space of the original building. This page: Interior view of the exhibition space. p. 65: View from under the new building's sculptural thatch roof.

对页：从室外看原建筑内的展览空间。本页：展览空间内景。第65页：新建筑雕塑般的芦草屋顶的下方。

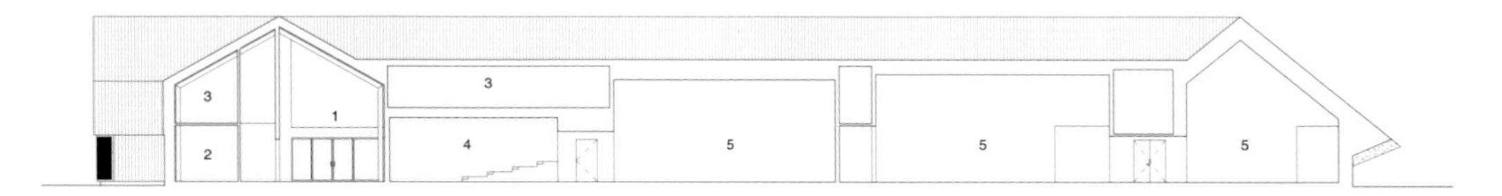

Section (scale: 1/500)／剖面图（比例：1/500）

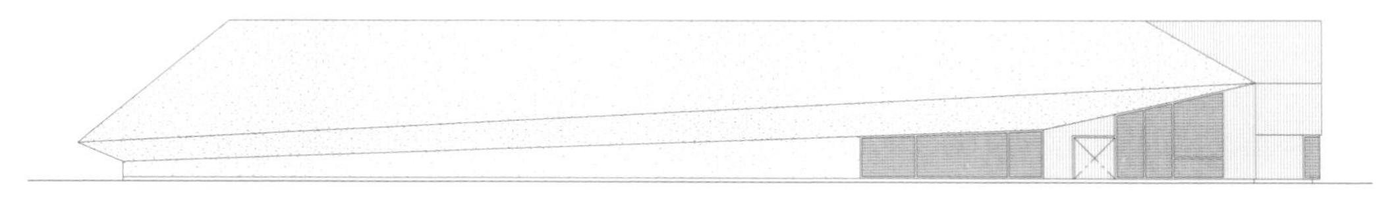

North elevation／北立面图

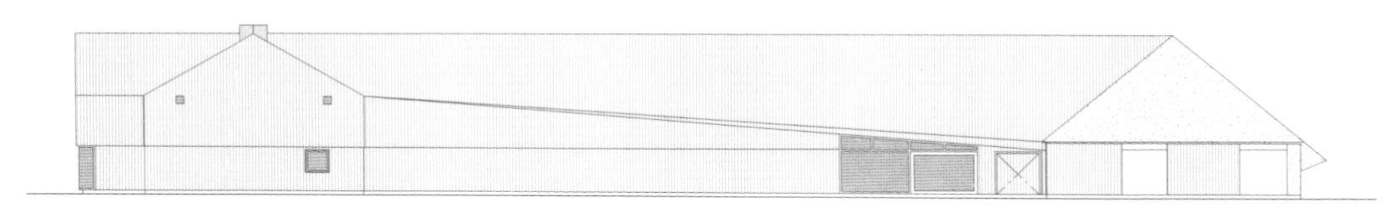

South elevation (scale: 1/500)／南立面图（比例：1/500）

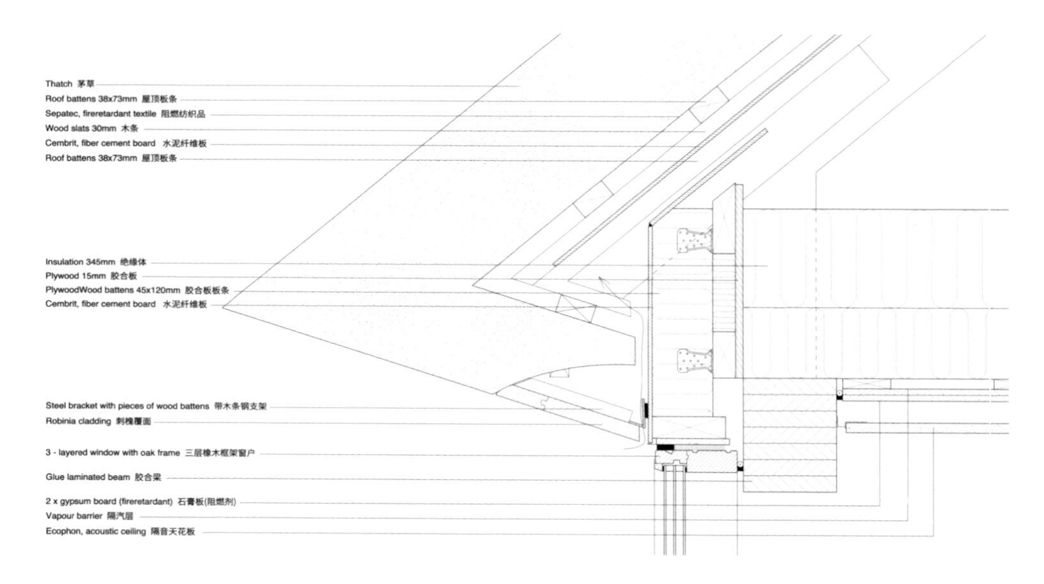

Detail section (scale: 1/15)／剖面详图（比例：1/15）

Dorte Mandrup

Trilateral Wadden Sea World Heritage Partnership Centre + Wadden Sea World Heritage Centre

Wilhelmshaven, Germany + Groningen, Netherlands 2018–

多特·曼德鲁普

瓦登海世界遗产三国合作中心+瓦登海世界遗产中心

荷兰，德国+格罗宁根，威廉港 2018-

A spectacular site

Located in the port city of Lauwersoog in the province of Groningen, the centre becomes another focal point for the Wadden Sea World Heritage Centre. The ambition is for the centre to become an internationally recognizable symbol highlighting the importance of protecting the unique natural World Heritage site which the Wadden Sea embodies.

The vast open landscape, the presence of the tide and above all, the 360 degree horizon makes this a remarkable place with a very strong identity. The centre is inspired by its location, in sitting at the intersection between the Lauwersmeer's calm waters in the harbor and the Wadden Sea where the tide dominates. The building will celebrate the meeting between different fields of knowledge and the movement of people and activities. It will foreground the presence of the most basic forces of nature, the sea and the tide.

The building itself draws on references from historic harbor piers, in resting on wooden stilts at the water's edge. Defined by open floor plates which create a continuous ramp climbing towards the sky, the structure highlights the vast surrounding horizon. The building is lifted from ground level providing full transparency, reiterating that it houses an inclusive space for everyone to enjoy, offering a place for experience, recreation and education. Outside the building a sloping plinth connects the roadside to the promenade level of the building and down into the harbor water. Both the plinth and floor plates of the building flow from the interior to the exterior, offering a meeting between the two.

Wadden Sea World Heritage Centre

A completely natural focus of the building is its'association with the surrounding water. We found inspiration for the gradual spiral-like incline of the floor from the changing tides of the Wadden Sea. Each level of the building leading to the next gives the guests a sense of being at one with the tide.

The building will house several exhibition spaces, office and research facilities, a seal centre as well as a café, restaurant and hotel. As people move up through the building, they will experience stunning 360 degree views of the Wadden Sea, the Lauwersmeer as well as the agricultural landscape in the distance. The restaurant is placed beside the seal centre giving guests views of the seals in the neighboring pool and a roof top terrace with panoramic views overlooks the World Heritage site.

Credits and Data
Project title: Trilateral Wadden Sea World Heritage Partnership Centre
Client: Trilateral Wadden Sea World Heritage Partnership Center Wilhelmshaven
Location: Wilhelmshaven, Germany
Design: 2018
Architect: Dorte Mandrup

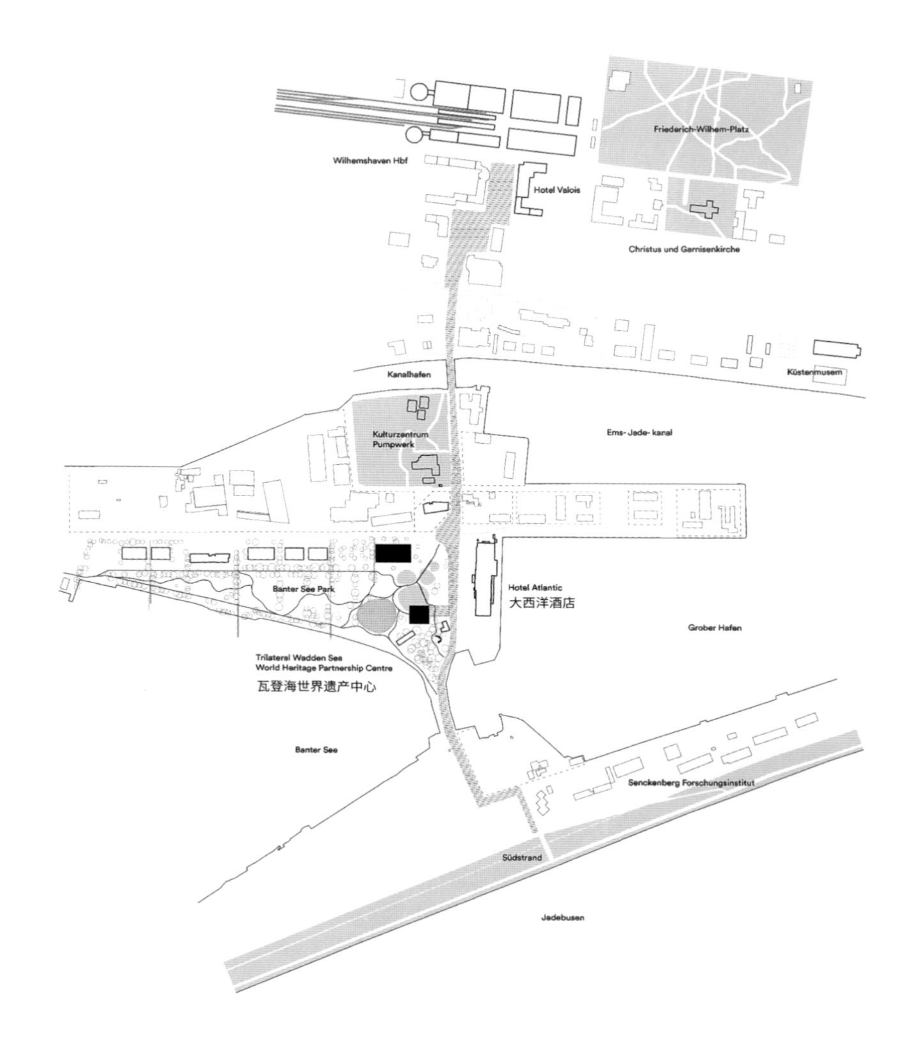

Site plan, Trilateral Wadden Sea World Heritage Partnership Centre (scale: 1/12,000)
总平面图，瓦登海世界遗产三国合作中心（比例：1/12,000）

p. 67: Trilateral Wadden Sea World Heritage Partnership Centre. The glass wall wraps the historic former military building. Images on pp. 66–73 courtesy of the Architect.

第67页：瓦登海世界遗产三国合作中心。历史悠久的原军用建筑被玻璃外墙包裹着。

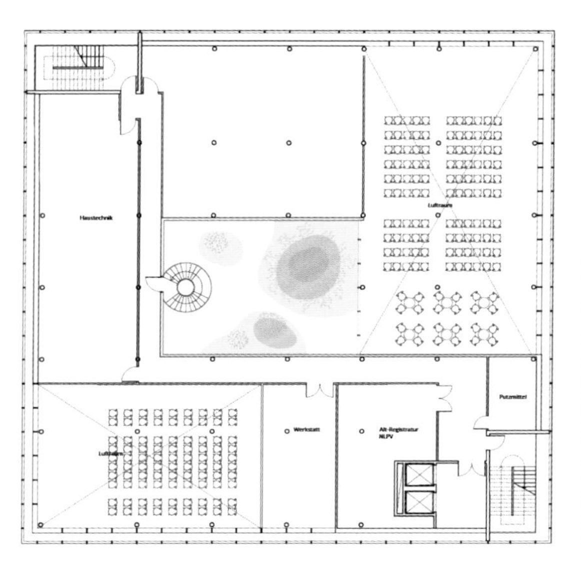

Level four plan／四层平面图

Level six plan／六层平面图

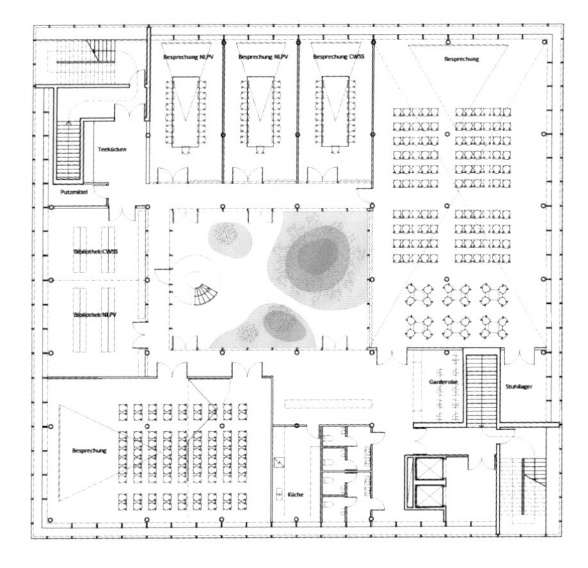

Level three plan／三层平面图

Level five plan／五层平面图

Plans of Trilateral Wadden Sea World Heritage Partnership Centre (scale: 1/450)
平面图，瓦登海世界遗产三国合作中心（比例：1/450）

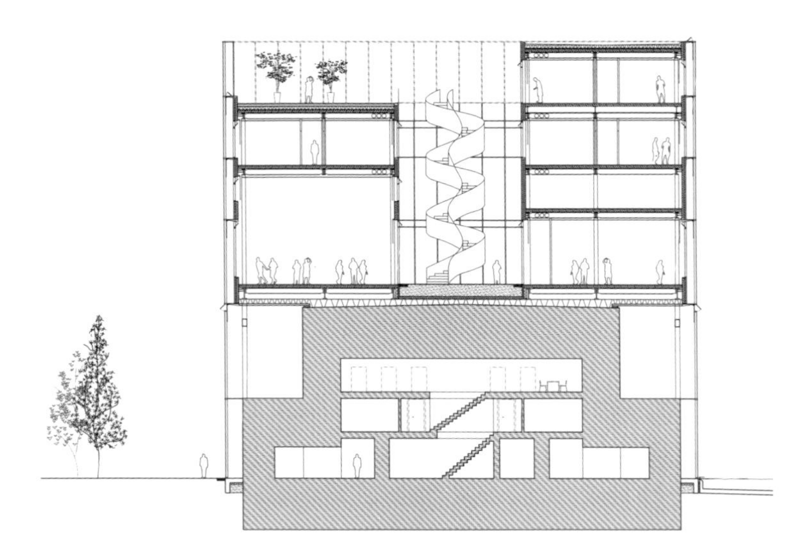

Section, Trilateral Wadden Sea World Heritage Partnership Centre (scale: 1/500)
剖面图，瓦登海世界遗产三国合作中心（比例：1/500）

壮观的场地

项目位于荷兰格罗宁根省的港口城市罗维苏格，将成为瓦登海自然保护方面的另一个焦点。瓦登海世界遗产中心的目标是强调保护这片独特而天然的世界遗产地区的重要性，并成为国际知名的标杆项目。

广阔的地形、潮汐的涨退以及最重要的360°环绕的地平线，都使这片场地极具个性而又令人难忘。受当地环境的启发，该中心被建在劳沃斯湖港口平静水域和波涛汹涌的瓦登海的交界处。建筑内将举办不同知识领域的集会，以及各种与人类息息相关的运动和活动。它将提醒人们不要忘记大海、潮汐这些大自然最根本的力量。

建筑本身参考了历史著名的港口码头，坐落在水边木制的支撑柱上。开放的楼板，形成了一条连续向上的倾斜步道，突出了周边延绵不绝的地平线。架空的结构创造了完全的通透性，重申了其中包括提供给大家休闲娱乐的空间，同时提供了体验、休闲娱乐和教育的场所。外部为斜向底座，连接着路边和建筑的海滨步道，并延伸进港口的水面之中。建筑的底座和楼板都从内延伸至外，这使内外环境得以交汇融合。

瓦登海世界遗产中心

这座浑然天成的建筑的着眼点在于它与周边水域间的联系。建筑从瓦登海变化的潮汐中获得灵感，设计了螺旋渐进状的楼板。从建筑各层到达下一层时，游客都有一种被卷入潮汐中的感觉。

建筑内部设有若干个展览空间、办公室和研究设施、一个海豹中心以及咖啡厅、餐厅和酒店。当人们拾级而上时，将会360°地体验瓦登海、劳沃斯湖和远方农业区令人震撼的美景。餐厅设在海豹中心一旁，用餐时客人们可以观赏旁边水池中的海豹，以及从屋顶露台俯瞰世界遗产的全景。

This page: Render perspective of the Wadden Sea World Heritage Centre. The design is defined by open floorplates and continuous ramp, supported by structures that reflect the historic harbour piers.

本页：瓦登海世界遗产中心的效果图。建筑采用了开放的楼板和连续的倾斜步道的设计，支撑结构的造型来源于历史著名的港口码头。

Site plan, Wadden Sea World Heritage Centre ／总平面示意图，瓦登海世界遗产中心

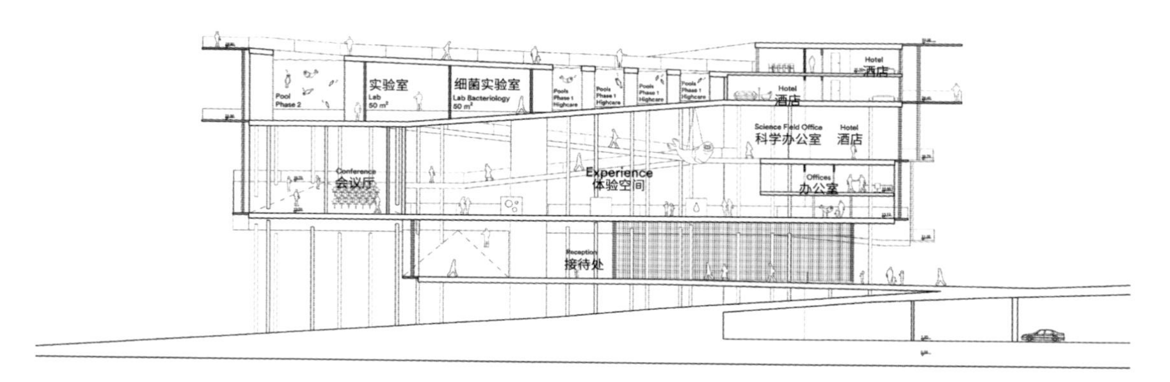

Long section, Wadden Sea World Heritage Centre ／纵剖面图，瓦登海世界遗产中心

Credits and Data
Project title: Wadden Sea World Heritage Centre
Client: UNESCO
Location: Groningen, Netherlands
Design: 2018 –
Architect: Dorte Mandrup

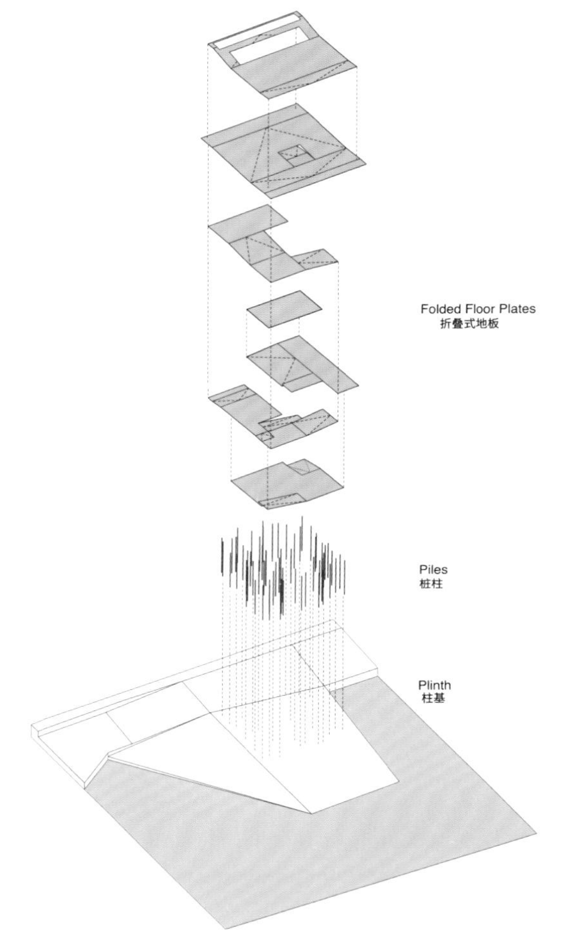

Exploded structural diagram, Wadden Sea World Heritage Centre
结构分解图，瓦登海世界遗产中心

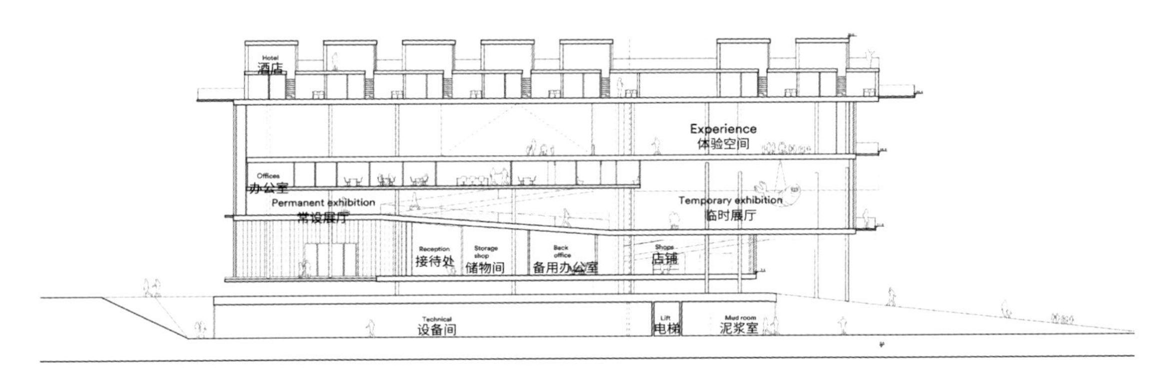

Short section, Wadden Sea World Heritage Centre／横剖面图，瓦登海世界遗产中心

Interview: Dorte Mandrup

Icefjord Center and Sustainability in Denmark

采访：多特·曼德鲁普

冰湾中心与丹麦的可持续发展

***a+u*:** Thank you for taking the time to do this interview with us. It is our pleasure to have you with us today. We understand that most of your work relates to sustainability issues, perhaps we could start with what sustainability is to you, in reference to one of your recent works in Ilulissat, the Icefjord Centre?

Dorte Mandrup (DM): The main focus for our work is not sustainability, but artistic interpretation of the given conditions. But sustainability is a natural consideration for all architects in Denmark. We have since the energy crisis in 1973 been very aware of energy consumption. There is almost no fossil energy resources in Denmark apart from a little natural gas, and the risks from nuclear energy was not accepted by the majority of people, which means that we have no nuclear power plants. The development and interest in clean energy like wind and solar power has been taking of from the 70 ties, and naturally also the political interest in the saving of fossil energy, by having, compared to other countries, much stricter demands for heat insulation, breaking of cold bridges etc. Building codes regarding heat loss started in the 70 ties and these demands that have been increasing with new building codes every decade. Right now the general building code in Denmark actually requires all houses to be equivalent of passive house standard. What I am saying is that the considering and knowledge about energy consumption is actually part of the upbringing of every Danish architect.

There is of course other aspects than reducing the carbon footprint in the running of a building. To build with non toxic and renewable materials with a low carbon footprint and altogether lower the carbon footprint in the production and maintaining of the building. Social and economic sustainability is a more holistic approach taking the social and economic impact of a building to the context into consideration.

The Icefjord center is placed in an extreme climate 250-km north of the arctic circle. There is snow most of the year from September to June, and in terms of materials, in Greenland there are no materials readily available - you have no wood, no forest, nothing. All the building materials that we are using here have to be shipped, in the short period when the sea is ice free and open. It is shipped mainly from Denmark, Norway, or Sweden. It means that we need to be extremely careful with the amount of materials and the type of materials that we use. So, what we are doing in the case of the Icefjord Center is, prefabricating everything so that it can be shipped in containers and taking as little space as possible in each container, to cut down on transportation. Secondly, we will float most of the building above the terrain, blasting as few rocks as possible, to avoid damaging the site, and to ensure the large amount of meltwater to drain freely in the spring. Another challenge we have is that we have to build in a very short period of time. In Greenland, snow and ice are present most of the year and as a result, we only have 2-3 months to build. Shipping the materials and getting them in place in a short time is also very important in this project as we need to be able to close the house before the winter comes. Once built, we will use high energy efficiency glass with three layers, which helps to keep the heat in as much as possible, reducing the energy used on

a+u:感谢您抽出宝贵的时间接受我们的采访。很高兴今天能邀请到您。我们了解到您的大部分项目都与可持续性的问题有很深的关联，也许我们可以先从最近您在伊卢利萨特完成的“冰湾中心”来谈谈您对于可持续性的理解。

曼德鲁普:我们工作的重点不在于可持续性，而是在于指定条件下的艺术性诠释。不过，丹麦建筑师天生便会考虑可持续性这个课题。自1973年能源危机以来，我们就十分关注能源消耗。丹麦除了很少的天然气几乎没有化石能源资源，同时大多数人不接受核能带来的风险，所以我们也没有核电厂。自20世纪70年代以来，丹麦一直有着对风能和太阳能等清洁能源的发展和兴趣。自然而然地，也有了节省化石能源的政治倾向。与其他国家相比，我们对保温隔热、冷桥断裂的要求更加严格。有关散热的建筑规范始于20世纪70年代，每十年都会更新。目前，丹麦所施行的一般建筑规范实际上要求所有房屋都必须符合被动式房屋（起源于德国的节能住宅）标准。我想指出的是，每位丹麦建筑师成长的过程中都会关注能源消耗的知识和考量，这是他们必须具备的最低职业素养之一。

当然，除了减少建筑的碳排放量外，建筑节能还有其他方法。比如使用无毒的低碳可再生材料可以从整体上在建筑物的生产和维护中降低碳排放量。我们需要从社会和经济层面考虑这些持续性的问题，更要关注建筑所在基地的背景、地方文脉所带来的影响。

“冰湾中心”位于北极圈以北250公里的极端气候中。从9月到次年6月中的大部分时间都下雪。就材料而言，格陵兰岛没有现成的建筑材料，没有木材，没有森林，没有石头，一无所有。我们在这里使用的所有建筑材料都必须在海洋无冰的短暂时段中运输。它们主要从丹麦、挪威或瑞典航运而来。这意味着我们对使用材料的数量和类型必须充分考虑。因此，在“冰湾中心”的建造过程中，我们预制了全部建筑构件，以便可以将其装在集装箱中，并在每个集装箱中尽可能少地占用空间，以减少运输成本。其次，我们使建筑物的大部分建于地势之上，尽可能减少对岩石的破坏，以避免损坏场地，并确保大量的融水可以在春季自由排出。我们面临的另一个挑战是，必须在很短的时间内建成这幢建筑。在格陵兰岛，一年中的大部分时间都有雪和冰，因此，我们只有2~3个月的时间来建造。在这个项目中，运输材料并在短时间内将它们放置到位也非常重要，因为我们需要在雪季真正到来之前完成建造。建成后，我们将使用三层高能效玻璃做窗，这有助于尽可能地保持热量，减少热损耗，同时控制用于暖气的能源使用。屋顶和地板采用50厘米厚的隔热层。对于表皮覆层，我们将使用一种不需要任何化学处理的木材，称为刺槐。这是一种在欧洲生长的硬木，非常耐水，因此可以不加处理，自然地使用。

a+u:我们谈论了建筑的材料，那么“冰湾中心”的形状是如何构想出来的？它也与您分享的那些可持续要素有关吗？

曼德鲁普:是的，对于“冰湾中心”来说，其形态是从不同层面考虑的结果。在恶劣的气候中，在大风和大雪的情况下，必须考虑积雪的问题，因此建筑需要被放置在积雪最少的位置。此外，建筑经过空气动力学计算而成的形态会使风吹向其立面，吹走积雪。关于雪的一个主要问题是风，风会不断地吹动雪堆；这意味着没有风时，才会导致建筑周围的大量积雪。

此建筑的外形还有另一个功能，即邀请人们走到屋顶上。它是人工与自然的连接通道。当你在建筑顶部行走

heating. The roof and floor are insulated with a 50 cm thick insulation and, for the cladding on the facade, we will use a type of wood that does not need to be treated by any chemicals, called Robinia. It is a hardwood grown in Europe. that is very resistant to water hence it can be left untreated and natural.

***a+u*:** While we are talking about the material aspect of the building, would you like to share with us how the shape of Icefjord Center was conceived? Does it also relate to those sustainable aspects you shared?

DM: Yes, for the Icefjord Center, the shape is a result of different aspects. In the harsh climate, with high wind speed and snow, you have to consider the snow build-up, therefore the building is placed to create the least snow build-up, and the aerodynamic shape allows the wind to blow towards the facade and carry the snow away, which is because one main problem of snow is the wind. It constantly relocates the snow, meaning when there is no wind the result is enormous amounts of snow pile-up. The form also plays another function by allowing people to walk up onto the roof of the building. It serves as a gateway between civilization and nature. So, as you walk on top of the building, you will start seeing this large Icefjord and as you move to the other side of the building, it continues into the wilderness. From inside the building, because it is cantilevered a little over the small valley, people are able to get a full view of the Icefjord that was previously blocked by a big rock. How the building responds to its surroundings are important, both in terms of its design and the sustainability aspects.

***a+u*:** Moving back to the last sustainability factor you mentioned earlier, on social sustainability, perhaps you could share with us how the Icefjord Center relates to this as well?

DM: First of all, the Icefjord Center is placed close to the icefjord which is a UNESCO protected site, for its unique nature. The Icefjord is starting by the world's largest active glacier at the edge of the icecap, and ends in the disco bay. Icebergs pack closely in the icefjord on their way to the disco bay. What has happened over the last 20 years is that the glacier has been visibly withdrawing. If you were to go up in a helicopter, the change is very apparent. The function of the building is exhibiting the importance of the icecap and climate change, so the interior is housing exhibition space, café, research library etc, and will be visited by the many tourists coming to Icefjord. At the same time the building is designed to be a local destination, to be as open, inclusive and inviting as possible. There is open covered public areas where you can rest, and the roof is a public space, where you don't have to pay an entrance fee. It is also a place where every year, people in Ilulissat can come to celebrate the return of the sun. Ilulissat is located above the Arctic Circle, meaning that every year the sun sets November 30th and does not return until midday January 12th. During these 40 min, the local community will gather to celebrate the return of the sun. So, by creating covered outdoor spaces and an open public roof, the Icefjord Centre is a great place for everyone in the community to come and celebrate.

***a+u*:** If sustainability is seen purely as something linked only with technology, energy efficiency and numbers, a person will not be able to tell if a house is sustainable just by looking at it. It does not have an expression. However, when we look into the streets, the buildings and houses in Northern Europe, it seems to tell a different story. What do you think? Is there something regional about sustainability?

DM: I think in the Nordic region, the driver of that would probably be to do with our tradition, the way we think of architecture. Our building culture is very much about inclusiveness and thinking about the public space. There is a tradition on thinking beyond that of the building. And, of course, we have Jan Gehl who has been teaching for many years and his books have had a strong influence, especially in Denmark, and before him a very strong functionalist housing tradition with among others Kay Fisker.

So, we stand on the shoulders of this tradition.

Right from the beginning, from being in architecture school, you are taught to understand how you can affect the space around you, and how you can create good public space. I think it is a cultural thing here that public spaces are as important as the

时，会渐渐看到这座巨大的冰峡湾，而当你慢慢移至另一侧时，视线会延伸进入旷野之中。从建筑内部看，由于它稍稍悬于小山谷之上，所以人们可以看到以前被一块大石头挡住的冰峡湾全景。无论在设计还是在可持续性方面，建筑物如何应对周围环境都很重要。

a+u：回到您先前提到的关于可持续性的最后一个因素，社会可持续性，您可否与我们分享一下“冰湾中心”是如何与此相关的？

曼德鲁普：首先，冰峡湾中心建在冰峡湾附近，这里因其独特的自然环境而受联合国教科文组织保护。冰峡湾始于世界最大活跃冰川的冰盖边缘，一直延伸到迪斯科湾，其间冰山在冰峡中紧紧层叠。但在过去的20年中，冰川已经明显消融后退。如果你乘坐直升飞机去看，变化是显而易见的。这个建筑的主要功能是暗示冰盖的重要性和气候变化，因此室内空间包含了展览空间、咖啡店、研究型图书馆等，以供众多来冰峡湾游览参观的游客使用。同时，该建筑的设计尽可能开放、包容并具有吸引力，以作为当地地标。人们可以免费使用有遮蔽的室外休憩空间和开放的公共屋顶。这里也是每年伊卢利萨特人欢庆太阳回归的地方。伊卢利萨特位于北极圈以北，意味着太阳从每年11月30日落山起，直到次年1月12日正午才会回归。在这40分钟内，当地社区居民将聚集在一起庆祝太阳的归来。因此，拥有有遮蔽的室外空间和开放公共屋顶的“冰湾中心”是社区居民庆祝该节日的好地方。

a+u：如果可持续性仅仅被视为与技术、能源效率和数字相关的事情，那么人们将无法只通过观看就分辨房屋是否是可持续的。它没有一个内容的外化形式。但是当我们看北欧的街道和房屋时，似乎并非如此。您对此怎么看？可持续性是否与区域性有关呢？

曼德鲁普：我认为在北欧地区，其驱动力可能与我们的传统以及对建筑的看法有关。我们的建筑文化非常注重包容性和对公共空间的思考。这种思考的传统不限于建筑。当然，我们有扬·盖尔，他教书育人多年，尤其在丹麦，其著作产生了深远的影响。在他之前，包括凯·菲斯克在内的一众学者与建筑师也为我们奠定了功能主义传统的基石。

因此，我们站在传统的肩膀上。

在学校学习建筑时，你从一开始就要学习理解如何通过设计影响周围空间以及如何创造良好的公共空间。我认为，公共空间与建筑本身同样重要，是文化性的。这样的讨论一直在进行中，只是我们没有将其标记为社会可持续性，而是简单地将其称为城镇规划等。另外，在丹麦语中，可持续性一词被写成“bæredygtighed”，这是一个很难翻译的词，其含义不仅仅是英语中的“可持续性”，它还具有持久性或长寿的含义。因此，在我们来看，它意指建造持久、永续的建筑。

a+u：您公司的设计是如何开始进行的？您首先考虑的事情之一是可持续性问题，还是从研究不同的设计表达入手？

曼德鲁普：我个人认为在建筑中，艺术的部分总是很重要的。可持续性是必要的，但不是唯一的目标。可持续性是必需，也是合理性的体现，合理的设计包括使用可持续材料，对建筑物进行适当的隔热等。这些都是建筑师智慧的一面，但并不是全部。我们的主要目标是通过项目做出艺术作品，在其中尽可能地探索艺术可能性。建筑的有趣之处在于，必须在回应功能的同时，构架并实现这些严格框架之外的最大潜力。一个好的建筑可以解决全部的这些问题。建筑设计就是在这些框架内工作时，尽可能多地回馈周围。在某种程度上，这也是可持续的。

a+u：和20年前相比，丹麦现在是否为鼓励可持续发展提供了一个很好的平台？您认为获得政府支持有多重要？

曼德鲁普：正如我在上面已经解释过的，丹麦一直特别注意能源消耗的问题。

现在，按照最新的建筑规范，几乎不可能在建筑立面中使用超过40%的玻璃，且使用的必须是三层节能玻璃。墙壁保温层也必须至少为30厘米厚。建造技术已经相当先进，可以最大限度地减少加热/冷却水的消耗并优化建筑废料管理，但仍有提高的空间。尽管年轻一代非常热心并且有意识地努力实现可持续生活，但拥有一个支持可持续发展并为此制定建筑法规等的政府同样非常重要。因为如果没有任何法规或建筑规范的要求，改变的速度将非常缓慢。以美国为例，有一些建筑师团队非常支持可持续性，但是

architectural piece itself. I feel that the discussion has always been there, only that we didn't label it as social sustainability, we simply called it town planning, and so forth. Also, in Danish, the term sustainable is written "bæredygtighed", which is a difficult word to translate and means a lot more than just 'sustainability' in English. It has a meaning about lasting, or longevity, and so the way we see it, is to build things that are made to last.

***a+u*:** How does your office go about designing? Are the sustainable issues one of the first things you draw upon, or do you begin with studying different design expressions?

DM: Well, I think personally, the artistic part of architecture is always important. Sustainability is necessary but it is not the only goal. It is something you do and it is part of being sensible, to design sensibly by using sustainable materials, properly insulating your buildings and so forth. But, that is not the main goal. The main goal is the artistic work we do with the projects, to explore as much as possible the artistic possibility. What is interesting about architecture, is that you have to frame, and to achieve, the broadest potential out of these strict frames, while answering to functionality. Good architecture resolves all of these issues. It is about giving back as much as you can when working within these frames, and in a way this is also sustainable.

***a+u*:** Do you think the situation now in Denmark provides a good platform for encouraging sustainability compared to 20 years ago? How important do you think it is to have the support of the government?

DM: Well as explained, Denmark has been quite aware of their energy consumption in particular.

Now, with the latest building code. It is almost impossible to use more than 40% glass in the facade, and the glass used will be a 3-layer energy efficient glass. Insulation of the walls also has to be at least 30 cm thick. The technical installations are pretty advanced to minimize consumption in heating/ cooling water use and optimize waste management, but there is still a lot to be done. Although the young generation is very enthusiastic and aware of trying to live sustainable, it is very important to have a government that supports sustainability, to make building codes and so forth. Because if there are no regulations or building codes that insist upon it, I think the change will be very slow. Take the United States for example, there are some groups of architects that are very supportive of sustainability, however, their building codes do not help them very much, and in the end most of what is being built there is really not very sustainable regarding material use, or energy consumption, in the end the client has the last word.

***a+u*:** In the struggle for sustainable architecture, architects are not the most powerful people, always finding themselves in the hands of clients and the economic situation. How do you think the role as an architect should be today?

DM: If we talk about social sustainability, architects shape physical surroundings, and public space is a big part of our responsibility. We need to be sensitive in making public spaces that are inclusive and inviting, creating a frame for social life. Although we hope to have responsible clients that share the same ambition of trying to make a sustainable building, it is also an architects' responsibility to make the client aware of what the possibilities are. We can find ways of addressing the space outside of the building in a responsible way and convince the clients that by giving back to the site, we can continue to gain from the site too. Also, I think we need to understand that architects cannot change the world, but we can play a part in making a new political agenda since we have been experts on the physical surroundings. In Copenhagen, especially in new areas, building codes and regulations require landowners to address the public spaces. So, I think it is very important to address that sustainability and energy efficiency is pretty advanced in the Nordic countries because there is a political world in which to do it, and not because the architects are more responsible in that sense.

他们的建筑法规对此并没有规定，导致大多数建造完成的建筑在材料、能源消耗方面其实并不符合可持续性原则，到最后，业主才是那个拥有决定权的人。

a+u：在为可持续建筑而奋斗的过程中，建筑师并不是最有话语权的人，他们总是会被业主和当时的经济形势左右。您如何看待今天的建筑师的角色？

曼德鲁普：如果我们讨论社会可持续性，那么塑造公共空间和周边环境便是建筑师的重要责任之一。我们在设计有包容性和吸引力的公共空间时要保持敏感，争取为公共生活创造一个良好框架。尽管我们希望能够遇到富有责任心的业主，与我们一同打造可持续建筑，但建筑师同样有责任让业主意识到可持续建筑的可能性。我们可以找到很多方法来负责任地处理建筑之外的空间，也可以说服业主通过回馈地方，同样能够从中持续获益。另外，我认为人们需要认识到建筑师无法改变世界，但我们作为物理环境方面的专家可以在制定新的政治规划中发挥一定作用。在哥本哈根，尤其是在新区，建筑法规要求土地所有者对公共空间进行考量。因此，我认为北欧国家的可持续发展和能源效率的高度先进性在于其政治环境，而并非建筑师在这件事上更具责任感。

Dorte Mandrup began her career with Henning Larsen architects and subsequently established her own drawing room in 1999. Dorte Mandrup herself is heavily involved in all projects. The studio consists of 60 employees from more than 10 different countries. Dorte Mandrup's work specializes in cultural and landmark buildings, adaptive reuse buildings, multifunctional buildings as well as in creating education and work spaces. Dorte Mandrup is honorary professor at the Royal Danish Academy of Fine Arts, School of Architecture, a member of the Louisiana Museum of Modern Art and will be a guest professor at Cornell University in the Autumn of 2018.

多特·曼德鲁普的职业生涯始于亨宁·拉森建筑师事务所，她于1999年建立了自己的工作室。工作室由来自十多个国家的60名员工组成，曼德鲁普本人深度参与了所有项目。曼德鲁普的作品专门研究文化和地标性建筑、自适应复用建筑、多功能建筑以及创造教育和工作空间。曼德鲁普是丹麦皇家美术学院建筑学院的名誉教授，路易斯安那现代艺术博物馆的成员，并在2018年秋季成为康奈尔大学的客座教授。

Essay:
Heirloom
Beate Hølmebakk

论文：
传家之宝
比阿特·霍尔梅巴克

The Oslo School of Architecture (AHO) in the 1980s: A school of around 180 students and a faculty of a couple of engineers and approximately twenty architects, almost all of them practicing.

Norway before the oil boom.
Building before the exit agendas.

There were three teachers we talked about: Wenche Selmer, Christian Norberg-Schulz and Sverre Fehn.

Since the mid 50s Wenche Selmer had designed houses that manifested a deep understanding of the qualities of domestic life. Her buildings were modest masterpieces. She possessed a fine-tuned sensitivity about a site's given conditions and profound knowledge of timber construction. Her freehand sketches were always in exact scale. Wenche Selmer's authority came from experience and sincerity. She spoke quietly about aspects of architectural practice that we understood were significant: About the importance of not placing a building on the site's best spot, about the centimeters that differentiate a good entrance from a bad one and about the relationship between beauty and logical use of materials, in line with their integral properties. She was a role model not only for the female students. She was part of the faculty at AHO for twelve years, from 1976 to 1988; she was never made professor.

Christian Norberg Schulz was a legend. In a school where, architectural theory was neither on the agenda nor on the time tables of the students he represented an intellectuality that we were both foreign to and proud of. We knew that he had collaborated with architect Arne Korsmo on the extraordinary row houses in Planetveien in Oslo, but at AHO he was an academic and his peers were international scholars. He introduced us to the history of the architectural universe, however, it was his interest in the world that surrounded us – the forests, the mountains, the snow, the valleys and the water – that made us understand that these (to us) natural and everyday phenomenon had a deeper and more existential meaning. He was a great photographer and an intense lecturer. From the auditorium in the basement of St. Olavs gate – accompanied by his own black and white photographs of ionic orders, baroque interiors and medieval timber – he insisted that we must understand the character of our places.

This strong call to search for the inherent quality of things permeated the school. We felt that to build anything meaningful we needed to get in touch with the essence of not only the site and the materials, but also of the program itself: What was a House, a Library, a Church? It was this pursuit for the elemental that in 1986, as a student in Sverre Fehn's Bygg 3 studio, led me to disregard the given church program, find a site on a small island in the southern archipelago and reduce the project to a simple chapel. Almost 25 years later Fehn chose the same place for his Chapel in Olavsundet, Ny-Hellesund (See pp. 86–93). A project commissioned by a group of local artists preparing the celebration of a new millennium.

Sverre Fehn was the master. It was as if the fact that he had not built so much made his authority larger and his work more impressive: A proof of his architecture being radical. We knew

1980年代的奥斯陆建筑学院(AHO)大约有180名学生,教职员工由一群工程师和大约20名建筑师组成,几乎所有人都在进行建筑实践。

石油繁荣前的挪威。

房地产投资浪潮前的建筑。

我们谈及了三位老师:温彻·塞尔默、克里斯蒂安·诺伯格·舒尔茨和斯维勒·费恩。

从20世纪50年代中期开始,温彻·塞尔默设计的房屋就表现出对家庭生活品质的深刻理解。她的建筑是低调的杰作。她对场地的既定条件有极高的敏感,并且精通木构。她徒手绘制的草图总是比例精确。塞尔默的权威来自于经验和真诚。她平静地谈到关于建筑实践一些极为重要的方面,例如不将建筑物放置在场地中最佳位置的重要性,通过几公分的差别区分一个好入口和坏入口,以及关于美与材料使用的逻辑性与协调性,以符合其整体性能。从1976年到1988年,她在AHO任教十二年,是一个优秀的榜样而不仅仅是女学生的榜样,不过她始终未能评上教授。

克里斯蒂安·诺伯格·舒尔茨是一位传奇人物。他在一所建筑理论既不是主要任务,也不在学生课程表中的学校里,展现了一种令人感到既陌生又钦佩的睿智。我们知道他曾与建筑师阿恩·科尔斯莫在奥斯陆的"星球方式"——这一非凡的联排住宅项目中合作,但在AHO他是一名学者,他的同行也是世界各地的学者们。他向我们介绍了建筑史。同时,正是他对周遭世界,如森林、高山、积雪、山谷和水的兴趣,使我们理解这些日常而自然的现象对我们具有更深刻的存在意义。他是一位伟大的摄影师,也是一位热情的讲师。在圣奥拉夫斯门地下室的礼堂里,伴随着他所摄的爱奥尼柱式、巴洛克式的室内装修以及中世纪的木材的黑白照片,他坚称我们必须了解所在地域的特点。

这种探求事物内在的强烈呼声响彻学校。我们认为要构建任何有意义的东西不仅需要接触场地和材料的本质,还需要触及功能类型:到底什么是住宅、图书馆、教堂?正是出于对最基本概念的追求,在1986年,作为一名学生在斯维勒·费恩Bygg (=建筑) 3工作室实习时,我舍弃了既定教堂项目,在南部群岛的一座小岛上找到了一块场地,将该教堂项目简化成一个小礼拜堂。约25年后,费恩在尼赫勒松的奥拉夫松德为他的礼拜堂选择了同一个场地(第86-93页)。这是由一群当地艺术家为庆祝千禧年而委托的项目。

斯维勒·费恩是一位大师。他建成的作品较少,但这令他的名望更盛,同时也令其作品更令人印象深刻。这也证明了他的建筑是激进的。我们知道在项目中回归本真是多么困难,并且开始意识到建筑整体性和商业成功之间可能存在着相反的依存关系。假设他的作品名录很长,或者他的业主很多,我们可能就不会相信他作品的价值了。但这些从来没有发生。通过研究他建筑作品的照片和清晰的建筑图纸可以追溯其设计的起源,即布鲁塞尔世界博览会挪威馆中的混凝土、木材和有机玻璃构成的自发约束构造(a+u 99:01);建造与生活之间的结合成就的威尼斯北欧馆永恒的空间(a+u 99:01);北雪平府(a+u 99:01)中在基础几何形态中不妥协的空间和结构组织;海德马克博物馆(a+u 99:01)混凝土路径通过光与暗影重述当地生活的故事。这些建筑都让我们深信不疑:它们是原创的。在斯维勒·费恩已建与未建的项目中,我们发现了意义和动机。

在圣奥拉夫斯门的一楼,斯维勒·费恩坐在书桌上或

how difficult it was to obtain essential qualities in a project, and we started to sense that there might be an inverted interdependency between architectural integrity and commercial success. Had the list of his built works been long, or his clients many, we might have mistrusted his work's value. But we never did. Studying the photos and the explicitly architectural drawings of his buildings we felt close to the origin: The self-imposed tectonic constraints of the concrete, timber and plexiglass construction in the Norwegian Pavilion for the World Exhibition in Brussels (a+u99:01). The marriage between built and living matter resulting in the timeless space of the Nordic Pavilion in Venice (a+u99:01). The uncompromising spatial and structural organization within the fundamental geometry of the Norrköping House (a+u99:01). The responsive concrete path retelling the story of local life through light and darkness in the Hedmark Museum (a+u99:01). These buildings left us in no doubt. This was original work. In Sverre Fehn's built and unbuilt projects we found meaning and motivation.

On the first floor of St.Olavs gate, sitting down at our desks or in front of the large blackboard, (a self-standing triptych still kept at the school as a relic), Sverre Fehn talked about death, virtue and love. Despite our profound respect for his work, we were not quite sure how to connect the fantastic, often humorous, fables unfolding along the fragile chalk lines on the black surface to the tangible qualities of his buildings and architectural drawings. We were young, he was in his sixtys. We were at the threshold of architecture, he was inside. We were looking for ways to go. He told us what he saw from where he was. He described a world where the act of building was a matter of life and death, where construction was an existential expression. Looking back, now being closer to his age, I see that what he did in his blackboard lessons and desk-tutorials was giving us a reason to believe in architecture, to understand that the doubt involved when choosing between the square and the circle, between vault or beam, brick or concrete, was a fundamental doubt, connected to our own existence. Thus, architects had a moral responsibility. Not being true, or honest, to the constructive idea meant not only a loss of architectural content, it meant a betrayal of conviction. Sverre Fehn's proposal for the Nordic pavilion at the Osaka World Fair in 1970 – the iconic breathing lungs – was not a politically and fashionably correct project in line with the ruling technological optimism. It was a harsh critique of the downsides of industrial development expressed through architectural means: A membrane structure evoking awareness of the air we depended upon. When in 2014 we were asked to recreate the unbuilt Osaka project (See pp. 106–117 inside the Ulltveit-Moe pavilion of the Architectural Museum (See pp. 118–125), Sverre Fehn's final project, the world had changed. A generation and a half after the Osaka World Fair, sustainability is now a concern of many, looking for truth in construction is not.

What have been largely lost during these decenniums are not the exposed materials, the explicit details, the expressive constructions or the fruitful involvement of nature into built form – features that often, and rightly so, are ascribed to the

大黑板（学校至今保留的，作为文化遗产的独立三联画）前，谈死亡、美德和爱。尽管我们对他的工作深表敬意，但我们不确定如何将他那随着精细的粉笔线稿而展开的、奇妙而幽默的寓言与他的建筑和图纸的切实品质联系起来。我们当时还年轻，他已年过花甲。我们刚踏入建筑的门槛，而他已经身在其中了。我们正在寻找道路，他告诉我们他从他所处的位置看到了什么。他描述了一个世界，在这个世界中，建筑表现生死攸关，建造是一种关乎存在的表达。我现在已经接近他当时的年龄，回想他在黑板前和课桌上所教授的，是给我们一个理由相信建筑，去理解在正方形和圆形，拱和梁，砖块和混凝土之间的选择，是与我们自身存在相关的根本问题。因此，建筑师负有道德责任。对建造思想不忠实，不仅是建筑内容的损失，更是对自己的信念的背叛。斯维勒·费恩在1970年大阪世界博览会上提出的北欧馆方案，具有标志性的“呼吸之肺”，并不是一个符合当时时代和政治的主流的技术乐观主义项目。它是通过建筑手段向工业发展的礼赞泼冷水：膜结构唤起了人们对其生存所需空气的关注。2014年，当我们被委托重建斯维勒·费恩的最后一个项目，位于建筑博物馆乌尔维特·莫展馆内（第106-117页）未建成的大阪项目（第118-125页）时，世界已经发生了变化。在大阪世博会之后近半个世纪的今天，可持续发展已成为大众的环境关注点，但在建造的一方依然不置可否。

在这几十年中，丢失最多的不是裸露的材料、精致的细节、富有表现力的结构或自然对建筑形式卓有成效的介入。这些特征时常且正确地归因于斯维勒·费恩的建筑。我们难以找到的是一种建筑与意义相关联的信念。建筑现在正在成为商品，而不是有存在价值的建构。在我们这个时代，斯维勒·费恩关于地平线、恐惧、屈服和欲望的描述只引起了部分人的共鸣，但除了他的作品、模型和图纸（这些都得到了很好的关注）之外，这些也是他的遗产，即一个建筑与基本人类体验密不可分的世界。

吉登达尔宫，2018年1月25日

我们聚集在斯维勒·费恩的出版公司总部大楼的中央空间，我的父亲在这里工作了40多年。家人、朋友和同事正从丹麦之家的仿制立面两侧进入这个宽敞的房间，这是费恩在他职业生涯即将结束时令人惊讶地放置在新的办公大楼之中的。我们在这里分享美食和回忆。这是一个温和的冬季下午，几小时前我们跟随父亲的棺木走到了坟墓。在谈话、敬酒和握手的间隙，我抬起头望向天窗，如今它已被雪覆盖一半，在混凝土表面产生了美丽的斑驳光影。我想到了生、死与建筑：冥冥之中，万物相连。

architecture descending from Sverre Fehn. What is hard to find is the conviction that architecture is related to meaning. Buildings are now becoming commodities rather than being constructions of existential content. In our time Sverre Fehn's tales of horizons, fear, caves and lust resonate only with a few, but apart from his built work, models and drawings – all well taken care of – this is his legacy: A universe where architecture is inseparably connected with fundamental human experience.

The Gyldendal House, January 25, 2018

We are gathered in the central space of Sverre Fehn's headquarter building for the publishing company where my father worked for more than forty years. Family, friends and colleagues are entering the large room from both sides of The Danish House, a replica of an existing facade on the site that Fehn, close to the end of his career, surprisingly placed inside the new office building. We are here to share a meal and memories. It is the afternoon of a mild winter day, only a few hours since we followed my father's coffin to his grave. Between the spoken words, the toasts and the handshakes I lift my eyes to the skylights that now, half covered in snow, produce a beautiful irregular pattern of light and shadow on the concrete surfaces. I think about life and death and architecture: It all connects.

Featured Norwegian Architects Profile:

Beate Hølmebakk is a graduate from The Oslo School of Architecture (1990). In 2004, together with Per Tamsen, they opened an architecture studio, Manthey Kula. The studio is published in monographs and in international magazines. Work by the studio has been exhibited in international architectural exhibitions and is part of the permanent collection of FRAC (France 2017). She holds a professorship at the Oslo School of Architect and Design.

Sverre Fehn (see p10, [13])

Atelier Oslo is an architectural office established in 2006. Founded by Nils Ole Bae Brandtzæg, Thomas Liu, Marius Mowe and Jonas Norsted. Atelier Oslo' s portfolio includes projects ranging from large cultural projects to single family houses and small installations. The development of each project focus on creating architecture of high quality in which the basic elements of architecture such as structure, materiality, light and space are particularly emphasized and reinterpreted in order to solve current challenges.

精选挪威建筑师简介：

比阿特·霍尔梅巴克毕业于奥斯陆建筑学院（1990）。2004年，她与佩尔·塔木森一起创立了建筑工作室曼蒂·库拉。工作室的作品被刊登于国际建筑杂志期刊，且曾在国际建筑展览上展出，并于2017年被法国当代艺术收藏FRAC永久收藏。她在奥斯陆建筑师与设计学院担任教授。

斯维勒·费恩 （见本书第9页，[13]）

奥斯陆建筑设计事务所始于2006年，由尼尔斯·奥莱·贝·布兰德兹格、托马斯·刘、马里乌斯·莫威和乔纳斯·诺斯泰德共同创立。事务所的作品从大型文化项目到单户住宅和小型装置。每个项目都着眼于创建高品质的建筑，其中特别强调并尝试重新解释建筑的基本元素，例如结构、材料、光线和空间，以解决当前面临的挑战。

Sverre Fehn

Chapel in Olavsundet

Sørlandet, Norway 1999

斯维勒·费恩
奥勒松礼拜堂
挪威，南挪威 1999

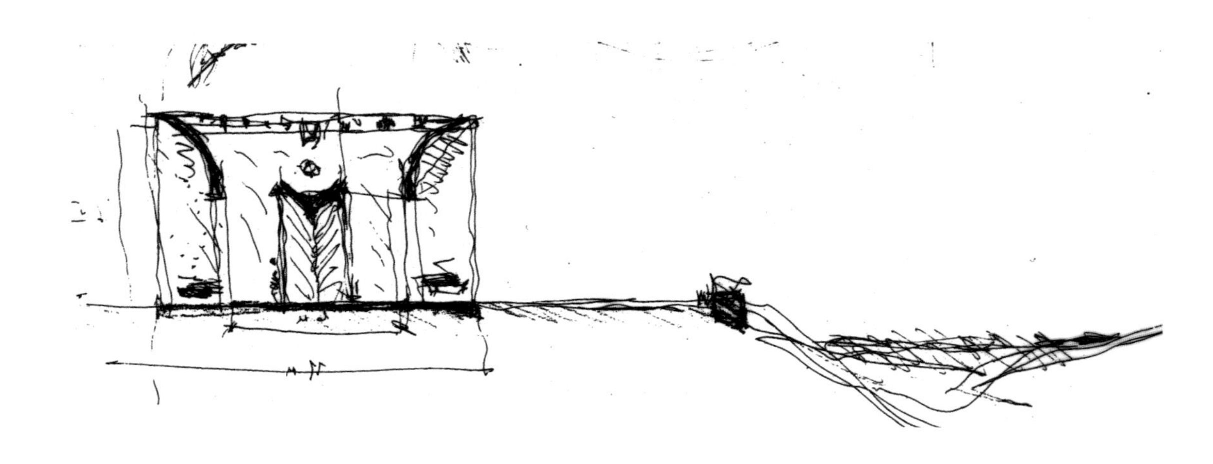

The project describes a simple form that will stand out in the great, dramatic landscape. The space will incorporate the four "corners of the world" with its large doors. The construction also creates four narrow openings in the space's diagonal directions. The sum of the openings will create distinct shading effects – it will seem like being in a room that demonstrates time. On the floor one can imagine a sundial, and the walls can imply the rhythm of the seasons. Externally the walls provide shelter via their concave forms.

In the structure's inner space and on the exterior, the chapel will stir the impulse to a meditative state. The structure puts the landscape on stage, and the light and the openings show fragments of the horizon. One becomes transfixed by the universal revealed by the sea.

The major walls would be executed in white concrete. The four doors set into the openings would be of wood, which can be alternately opened and closed as dictated by weather and wind. The visible wooden construction lends the ceiling a geometric form.

The four openings would be executed in thick glass in order to give the space the necessary silence.

Site plan (scale: 1/1,100) ／总平面图（比例：1/1,100）

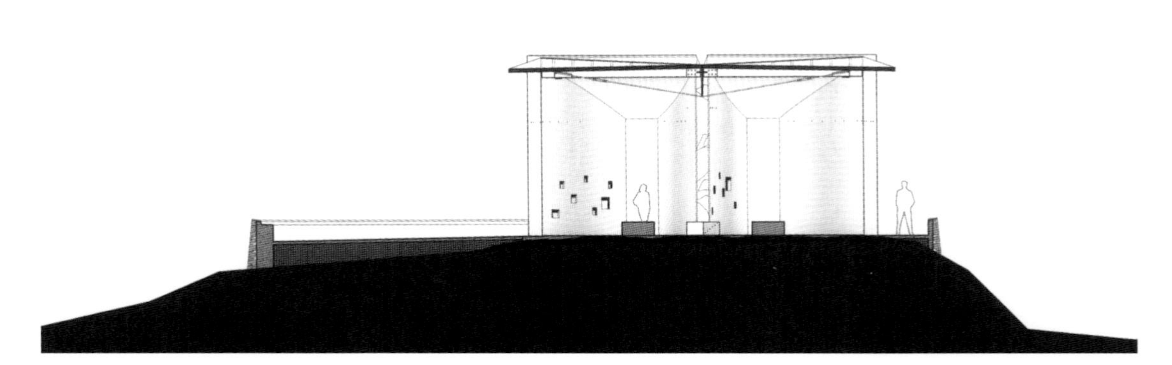

Section (scale: 1/400) ／剖面图（比例：1/400）

p. 86: Sectional sketch by the Architect. p. 87: Render view of the chapel from the south. Opposite, above: Render view of the interior. Opposite, below: Study model. Images on pp. 86–93 courtesy of the Architect.

第86页：剖面手绘图。第87页：从南侧看向礼拜堂的效果图。对页，上：室内效果图；对页，下：研究模型。

这个项目诠释了一种建筑形式，虽然简单，却可以在广阔且梦幻的大自然中独树一帜。建筑内的空间分别由4个带有大门的“世界的角落”构成。同时，结构上，建筑也在4个对角线的位置做了狭窄的开口，这些开口在室内产生了明显的光影效果，使内部空间仿佛有了显示时间的功能。站在室内地板上的人可以想象出一个日晷，而四周的墙体则暗示季节的变换。在建筑外部，墙体通过凹面的形状创造了一个抵御风吹日晒的场所。

从建筑的内部空间到外部造型，这座教堂都在尝试唤起人们进入冥想。建筑的构造将周围风景置于舞台上，而光线与开口则展示着地平线的片段。人们为大海所展现的普遍的浩瀚而深深折服。

主墙体计划由白色混凝土建造，而安装在开口处的4扇门则为木制，可以根据天气和风的状况自由开合。天花板上外露的木结构使整个顶部显现出几何学的造型。

4个开口被安装上厚玻璃，以确保内部空间的宁静。

This page: Visible wooden construction lends the ceiling a geometric form. pp. 92–93: Detail drawings.

本页：外露的木结构使天花板呈几何学造型。第92-93页：细部详图。

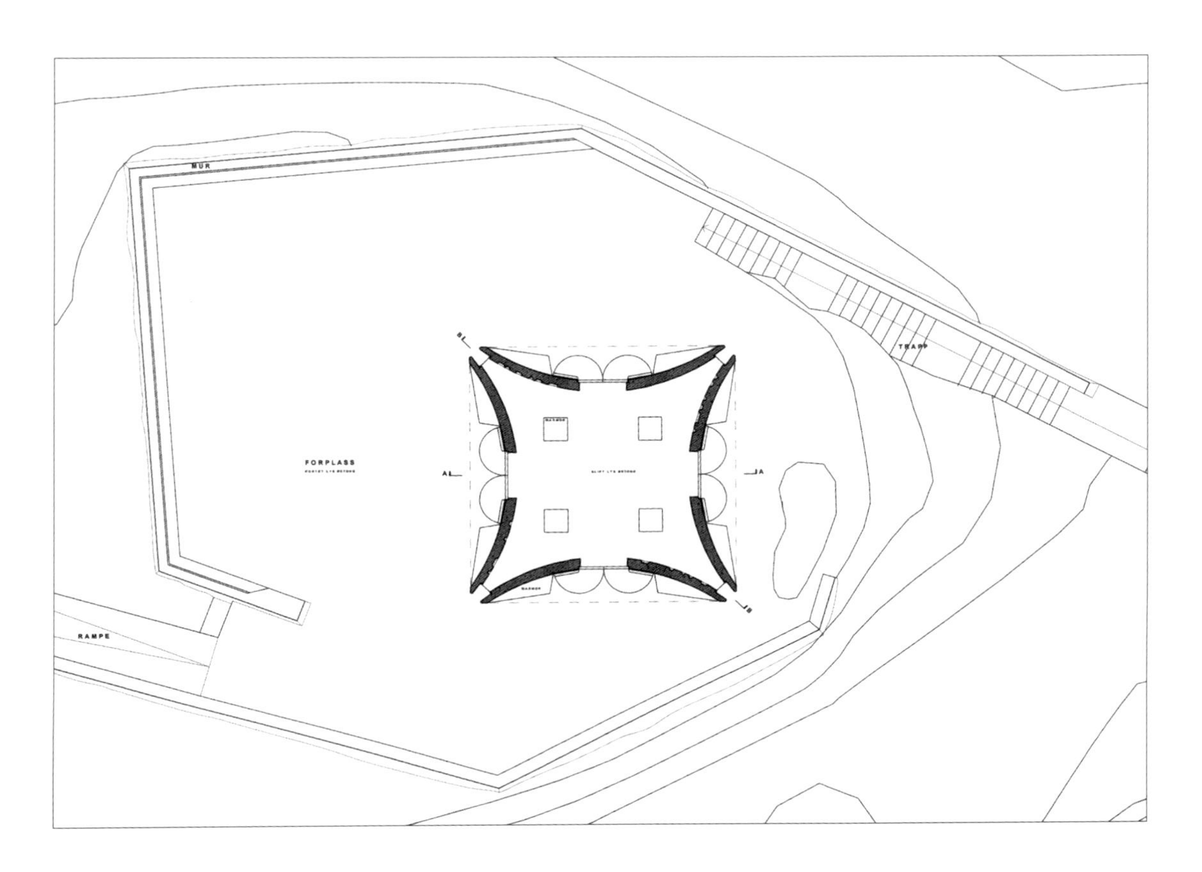

Plan (scale: 1/250) ／平面图（比例：1/250）

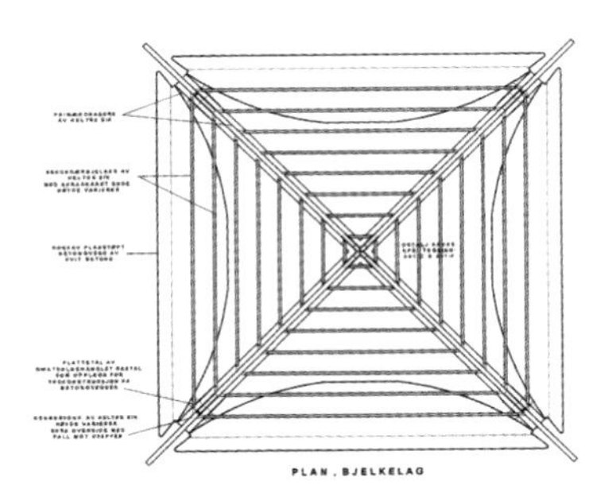

Ceiling plan (scale: 1/250) ／天花板平面图（比例：1/250）

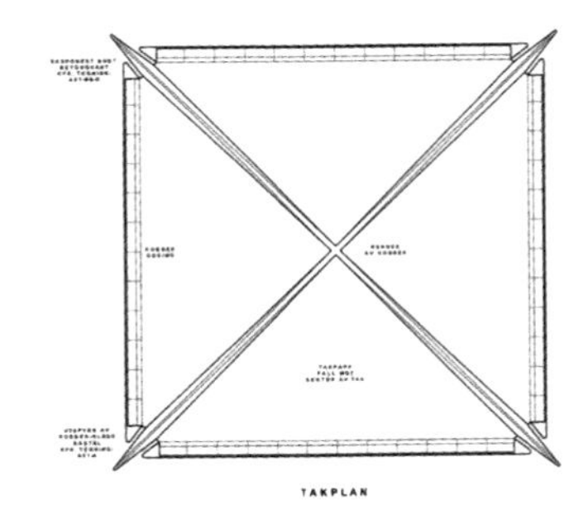

Roof plan ／屋顶平面图

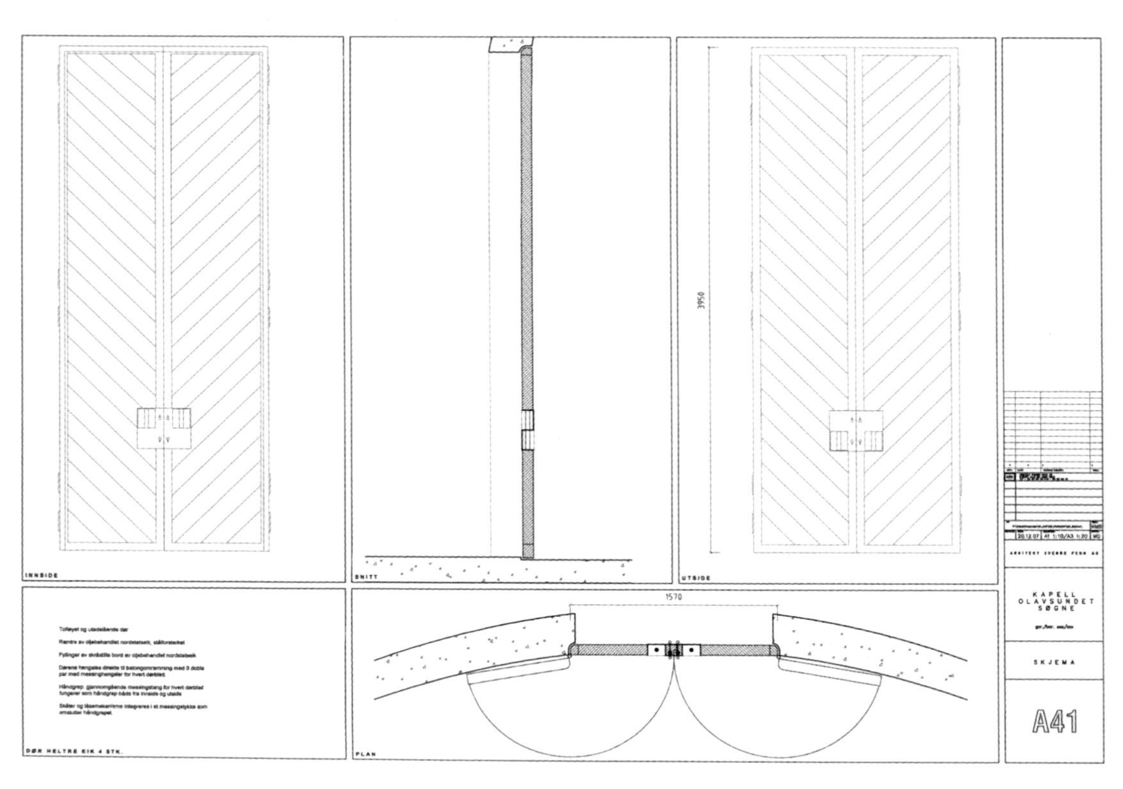
INNSIDE
SNITT
UTSIDE
3950
1570
PLAN
DØR HELTRE EIK 4 STK.
20.12.07
A1 1:10/A3 1:20
ARKITEKT SVERRE FEHN AS
KAPELL
OLAVSUNDET
SØGNE
SKJEMA
A41

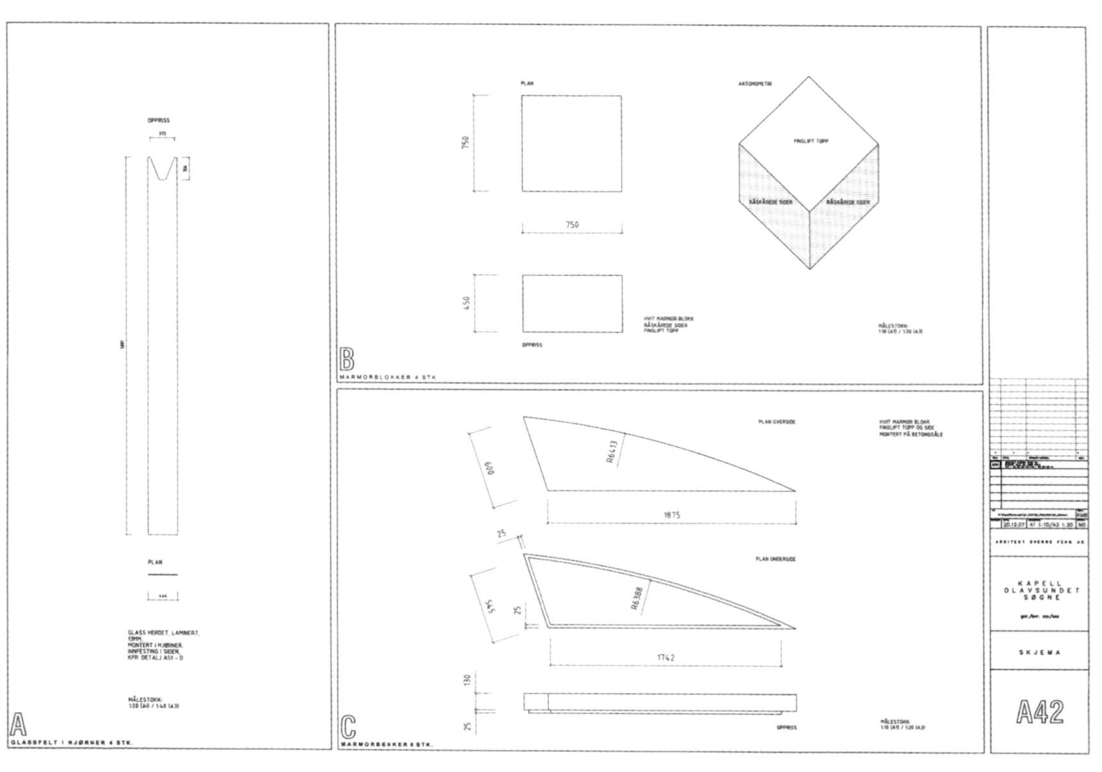
OPPRISS
PLAN
GLASS HERDET, LAMINERT,
MONTERT I HJØRNER,
INNFESTING I SIDEN,
KPR. DETALJ A51 - D
MÅLESTOKK:
A
GLASSFELT I HJØRNER 4 STK.
PLAN
AKSONOMETRI
FINSLIPT TOPP
BÅSKÅREDE SIDER
750
750
450
OPPRISS
HVIT MARMOR BLOKK
BÅSKÅREDE SIDER
FINSLIPT TOPP
MÅLESTOKK:
B
MARMORBLOKKER 4 STK.
PLAN OVERSIDE
HVIT MARMOR BLOKK
FINSLIPT TOPP OG SIDE
MONTERT PÅ BETONGSÅLE
600
R6413
1875
PLAN UNDERSIDE
25
545
25
R6388
1742
130
25
OPPRISS
MÅLESTOKK:
C
MARMORBENKER 8 STK.
20.12.07
A1 1:10/A3 1:20
ARKITEKT SVERRE FEHN AS
KAPELL
OLAVSUNDET
SØGNE
SKJEMA
A42

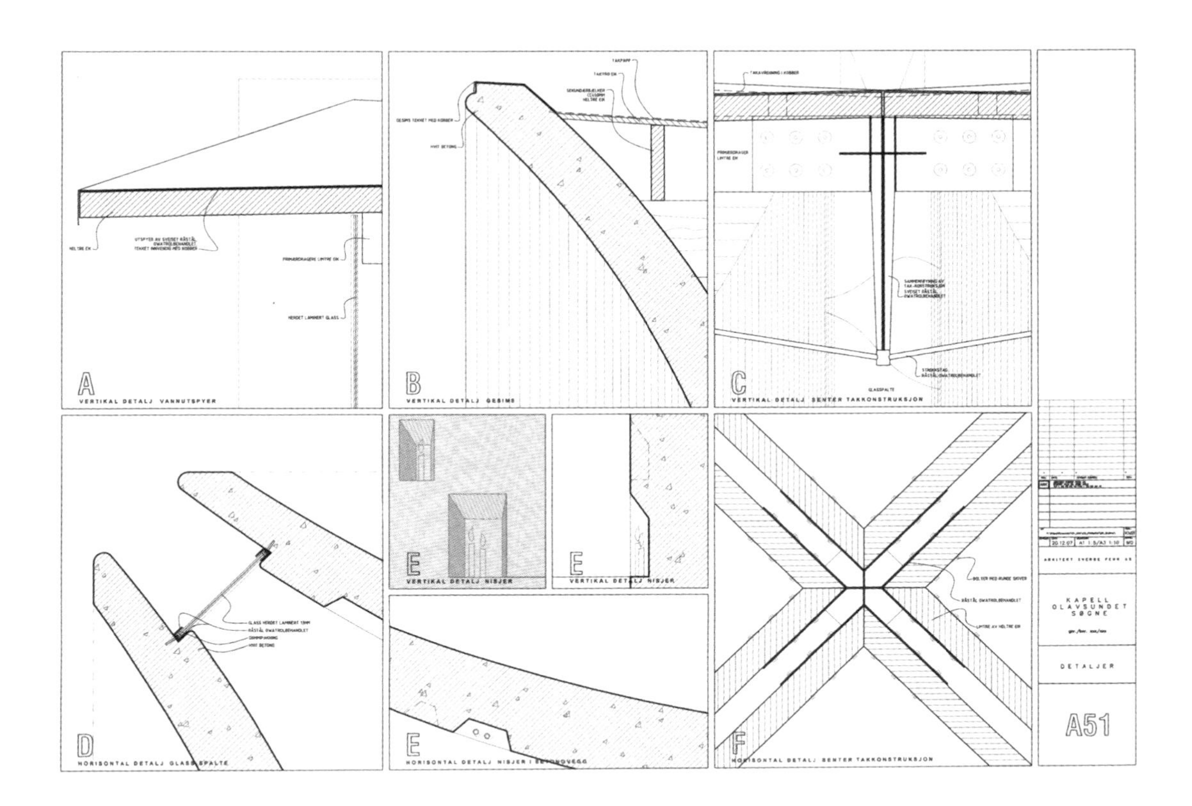

Credits and Data

Project title: Chapel in Olavsundet

Location: Ny-Hellesund, Sørlandet, Norway

Design: 1999

Architect: Sverre Fehn

Collaborator: Martin Dietrichson

Sverre Fehn

Gyldendal House

Oslo, Norway 2007

斯维勒·费恩

吉尔登达尔大厦

挪威，奥斯陆 2007

Norway's largest publishing house, Gyldendal, has been located in the block between Universitetsgata and Sehesteds plass since the beginning of the prior century. Behind the original 1800s-era facades a new headquarters has been constructed for the company: Gyldendal House.

Having resided in an inappropriate structure over a long period, Gyldendal wanted a modern and future-oriented building with a great degree of robustness. Additionally, they wanted the new building to be communicative and provide the publishing house with an identity derived from its architecture.

Sverre Fehn was engaged in 1995 to draft possible solutions based on the original structure. The historical context and the consistent architecture of this area of the city inspired the initial concept of maintaining the old facades toward the streets. Yet within, Gyldendal could have a new, specially tailored building with a clear organization and high architectonic quality. In 2004 detail work and construction began. 9,000 m² over five floors plus basement were completed in 2007.

The main entrance to Gyldendalhuset, with its legendary copper door, is at street level on Sehesteds plass. The ground level stretches through the entire block out towards Universitetsgata, providing one continuous urban space, where the reception area and the adjacent functions such as canteen, conference and informal meeting rooms encourage interaction between the publishers and the general public.

The building's main construction is of light grey concrete. The floor in the common areas and the permanent fixtures are of light oiled oak. The interplay between the plastic, monolithic concrete and the warm oak characterizes features prominently throughout the building.

Ibsenhallen (Ibsen Hall), centrally located within the new complex, is a large open space over five floors, with surrounding galleries and adjacent office areas. An overhead lighting construction of sculpturally-formed concrete pyramids carries a soft light through the space and into the building.

The open, floor-through structure provides a clear organization. The vertical communication, in the form of an arched concrete construction, lies centrally positioned between the reception area, Ibsenhallen and the open 'plaza' on each floor, with a small courtyard on each side and meeting rooms and quiet rooms. The office areas are effective and flexible, with technical installations laid under custom flooring.

The interplay between the large space, the office areas and the surrounding facades conveys an illuminated openness as well as an intimacy in the work spaces.

Hamsunsalen (The Hamsun Room), a sub-level auditorium by the reception area, and 'Danskehuset (Danish House)', an important historical reference point for Gyldendal, are among the other notable architectonic elements of the building.

This page: View from the atrium, with the main stair and service area. Photos by Dag Alveng.

本页：设有主楼梯和服务区域的中庭内景。

This page: View of the auditorium.

本页：讲堂内景。

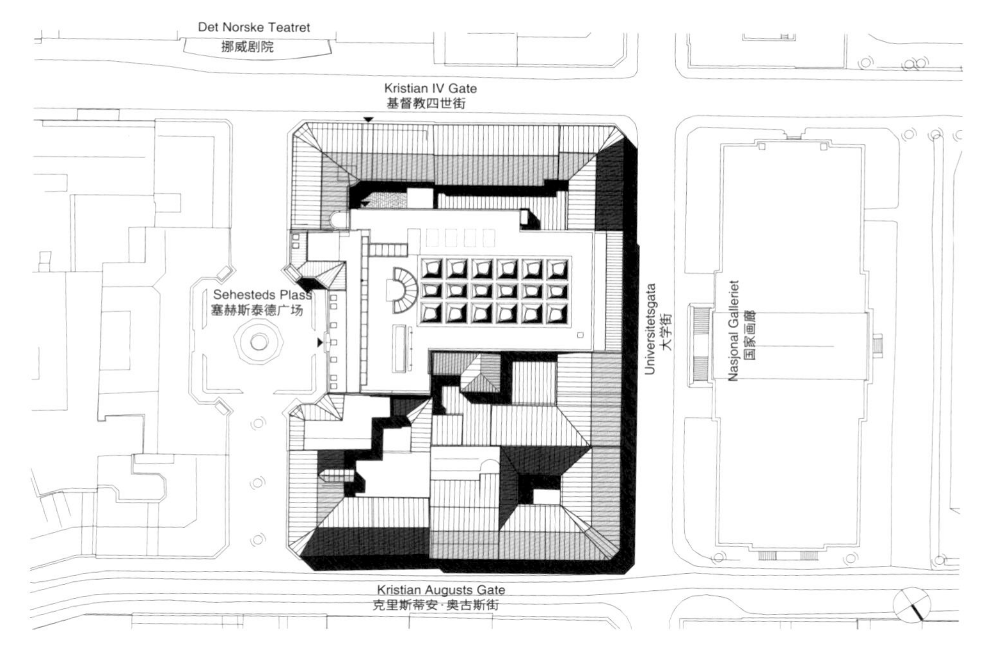

Site plan (scale: 1/1,500) ／总平面图（比例：1/1,500）

挪威最大的出版社，吉尔登达尔出版社，从上个世纪初以来一直坐落于大学街和塞赫斯泰德广场之间的街区。在19世纪的原有立面的后面，这座名为“吉尔登达尔大厦”的公司新总部被建造了起来。

由于已经在一座使用不便的建筑内办公了很久，公司方面希望新总部是一座现代而面向未来且经久耐用的建筑。此外，他们还希望新的建筑拥有良好的沟通氛围，并能从建筑上就看出这是一座具有出版社独有气质的建筑。

斯维勒·费恩1995年被委托参与这个项目，一起探索基于原有结构的设计解决方案。在受到这一地区历史文脉以及周边建筑协调性的启发后，建筑师得出了保留面朝街道古旧立面的原始提案，即初期的概念来源。但在建筑内部，吉尔登达尔可以拥有一个结构明晰且高品质的、崭新的、量身定制的建筑空间。详细的作业和施工开始于2004年，到2007年，包括地上5层和地下部分总计超过9,000平方米的建筑全部施工完成。

吉尔登达尔大厦的主入口配有古老的铜质大门，位于面朝塞赫斯泰德广场的地上一层。建筑的一层面向大学街，并延伸至整个街区。接待区以及相邻的食堂、正式和非正式的会议室等功能区的设计鼓励出版社员工与大众互动，从而创造了一个连续的城市空间。

建筑主体由浅灰色的混凝土构成。公共区域的地板和常设装备表面为清漆橡木。塑料、整体式混凝土和温暖的橡木之间的相互作用，营造出整座建筑鲜明的个性。

易卜生大堂是位于新建筑中心、通高5层的巨大开放空间，周围遍布长廊和办公区域。头顶的照明构造由混凝土制成，呈雕塑般的金字塔形，将为整个空间和建筑内部带来柔和的光线。

开放且跨越各层的构造赋予了建筑明快的空间组织形式。由混凝土构成的弧形竖向交通空间被设置在各层接待区、易卜生大堂以及开放的“广场”的中心部，其两旁有着小中庭、会议室和安静的房间。办公区域铺设定制的地板，并在地板下设有技术设备，使用起来高效而灵活。

大空间、办公区以及围合四周的立面相互作用，并通过灯光照明创造出了开放感和工作空间内的亲切感。

“汉姆生大厅”是一个位于接待区旁边的地下讲堂，而“丹麦屋”是吉尔登达尔重要的历史遗迹。此外，这座建筑中的其他元素也格外引人注目。

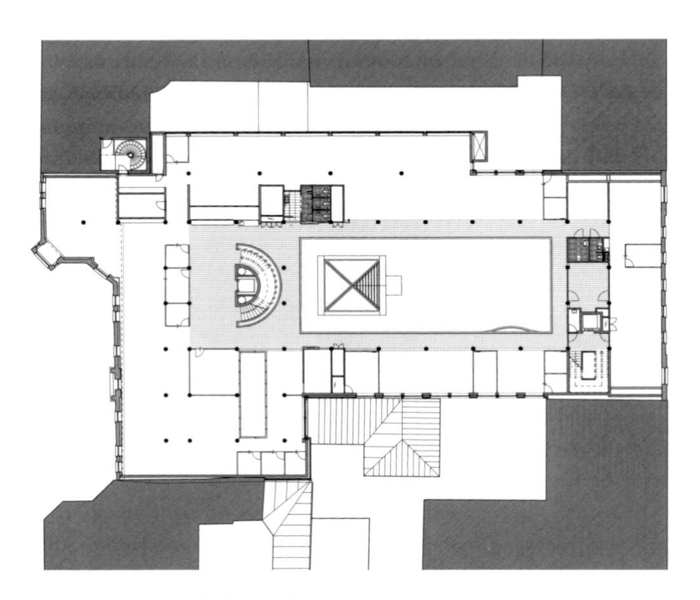

Typical floor plan ／标准层平面图

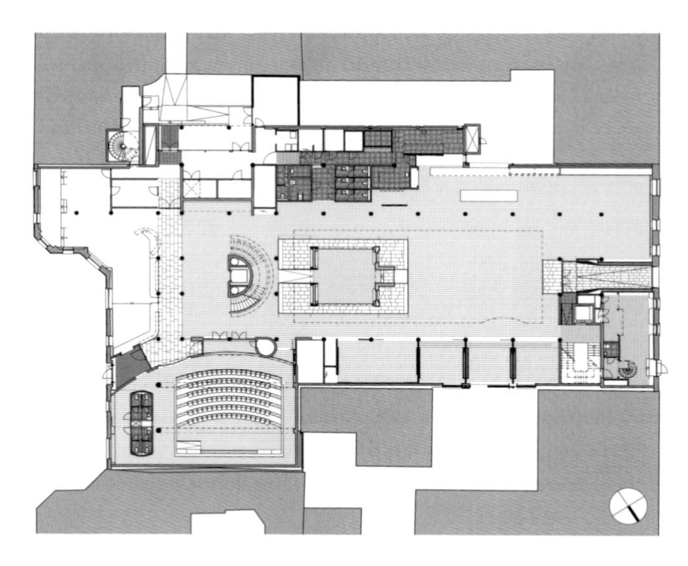

Ground floor plan (scale: 1/800) ／一层平面图（比例：1/800）

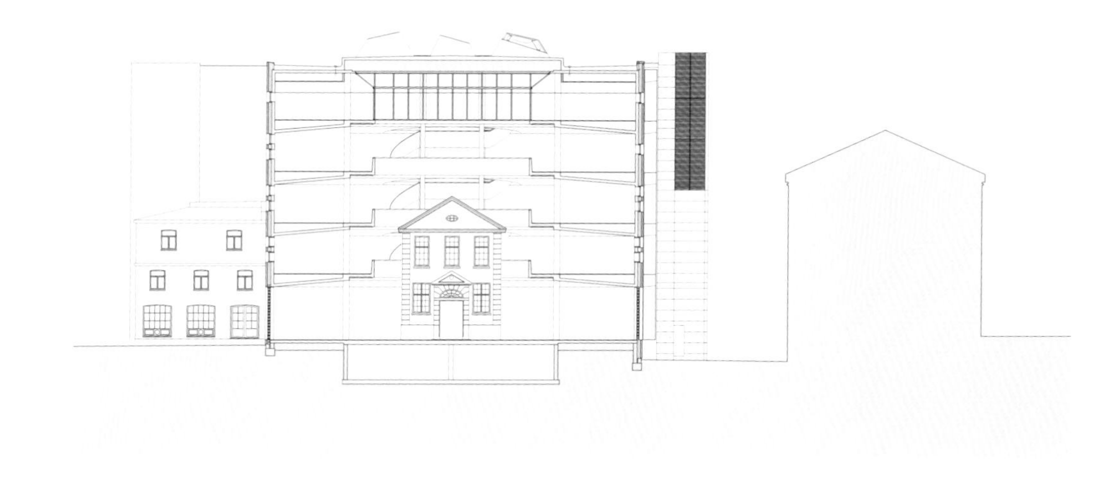

Short section ／横剖面图

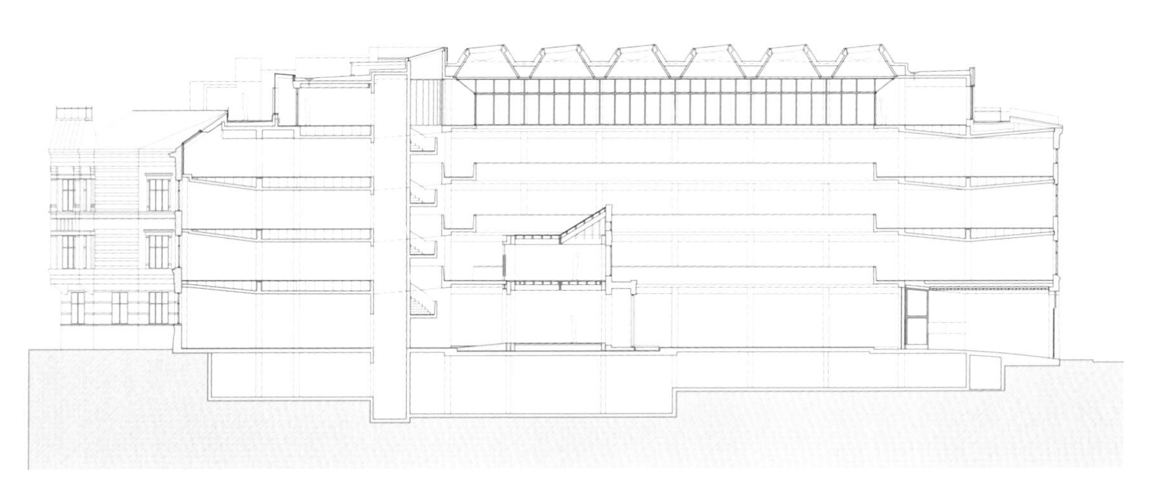

Long section (scale: 1/500) ／纵剖面图（比例：1/500）

p. 99: View from the atrium, with "The Danish House" in the middle. All photos on pp. 94–105 by Nils Petter Dale unless otherwise noted. Opposite: Glass brick wall providing light into an interior space. Photo by Dag Alveng.

第99页：中庭内景，正中间为"丹麦屋"。对页：光透过玻璃砖照进建筑内部。

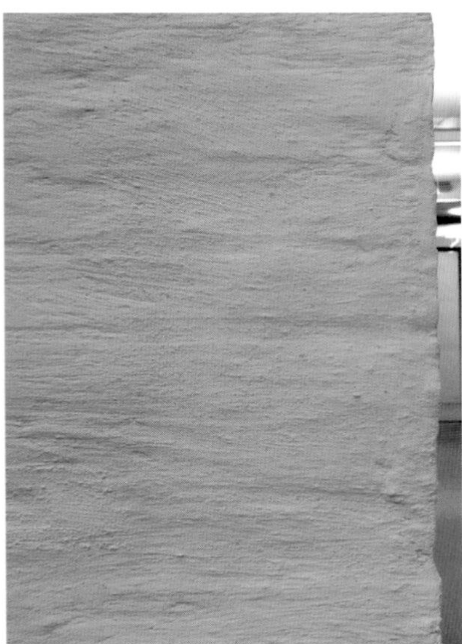

Credits and Data

Project title: Gyldendalhuset (Head office of Gyldendal ASA)
Client: Gylendal ASA
Program: Officebuilding
Location: Sehestedsgate 4, Oslo, Norway
Built: Opening 2008
Architect: Arkitekt Sverre Fehn AS
Project team: Inge Hareide (project leader), Kristoffer Moe Bøksle, Aleksander Wærsten, Halvor Kloster, Henrik Hille, Bård Hoff, Martin Dietrichson
Entrepreneur and project management: Vedal prosjekt AS
Engineers: RIB: Rambøll Norge, RIV: VVS- og KlimaRådgivning, RIE: Støltun
Project area: ca 9,000 m²
Cost: 190 mill NOK

This page: Close up views of the interior wall. Opposite: View of the atrium and gallery. p. 104: View between the main stairs and corridor. p. 105: Exterior view of the glass brick facade wall.

本页：室内墙近景。对页：中庭和回廊内景。第104页：主楼梯和回廊间的内景。第105页：玻璃砖墙的室外立面。

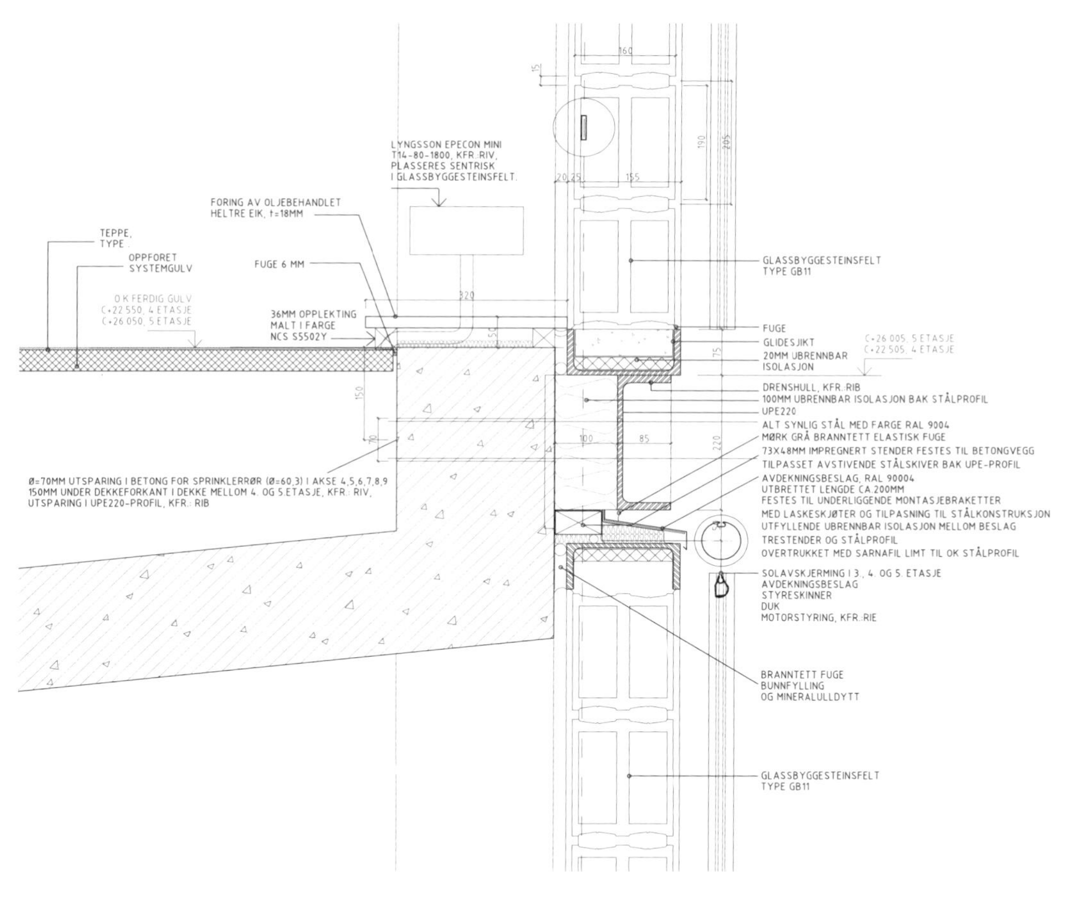

Detail section of the glass brick facade wall (scale: 1/12) ／玻璃砖外墙的细部详图（比例：1/12）

Sverre Fehn

The National Museum – Architecture

Oslo, Norway 2008

斯维勒·费恩

挪威国家建筑博物馆

挪威，奥斯陆　2008

Credits and Data

Project title: The National Museum - Architecture
Client: Norwegian Ministry of Culture & The National Museum
Program: Museum
Location: Bankplassen 3, Oslo, Norway
Built: Official opening 2008
Architect: Arkitekt Sverre Fehn AS
Project team: Martin Dietrichson (Project leader), Marius Mowe, Kristoffer Moe Bøksle, Inge Hareide, Henrik Hille
Project Management: Statsbygg
Entrepreneur: Atlant AS
Engineers: RIB: Dr. Techn. Kristoffer Apeland AS og Terje Orlien, RIV: VVS- og KlimaRådgivning AS, RIE: Støltun AS
Project area: ca. 3,800 m²
Cost: Grosch og Magasinbygget: 130 mill kr. NOK, Pavillion: 37,5 mill kr NOK Paid for in full by Jens Ulltveit Moe

pp. 106–107: Interior view of the pavilion supported by pillars bearing a delicate shell-shaped roof of light concrete. Photo by Candida Höfers. This page: General view of the old building and the new pavilion. Opposite: View of the new pavilion from Myntgata and Akershus Castle. All photos on pp. 106–117 by Nils Petter Dale unless otherwise noted.

第106-107页：展览馆内景。由柱子支撑的、轻型混凝土制成的精致壳形屋顶。本页：原有建筑全景和新建的展览馆。对页：从明加塔和阿克斯胡斯城堡处看新展览馆。

landskap

Norges Bank's first branch in Oslo, at Bankplassen 3, was designed by Christian Grosch. It was completed in 1830 as one of the country's first monumental buildings in the Empire Style. In 1910, Riksarkivet (The National Archival Services of Norway) took over the premises and expanded the site with a four-story archival wing. Since 1990 the building had stood unoccupied, scarred by several unfortunate renovations and a general lack of maintenance. Yet it also had protected status from Riksantikvaren (The Directorate for Cultural Heritage in Norway).

In 2001 the Norwegian Museum of Architecture submitted a proposal to the Ministry of Culture concerning relocation of the museum to Bankplassen 3. The proposal received much support in the media, and in April of the same year the government approved the proposal. The project was carried further after the Norwegian Museum of Architecture became a part of the National Museum of Art, Architecture and Design in 2003, and was completed in January 2008.

Sverre Fehn was the architect for the rehabilitation of both the old structure and the new exhibition Pavilion in the park toward the southwest. This Pavilion was financed by Jens Ulltveit-Moe, while the renovation of the Grosch building and the archival wing was underwritten by the Ministry of Culture.

The entrance to the museum is located at the original main entrance from Kongensgate. From this point one is led into the structure to where all the visitor-oriented functions are primarily situated at ground level. In the Grosch building there is a main hall with reception and bookstore, cafe, exhibition space, toilets and coat-checks. From the hall, visitors are directed to the Pavilion, a space for temporary exhibitions, and to the archival wing, where the museum's permanent exhibition is displayed. Toward the south there is access to the park with outdoor dining in the summer season. The handicapped and personnel entrance as well as the service entrance are on Nedre Slottsgate, while administration, library and assembly locations are located on the second floor of the Grosch building. In the two upper floors of the archival wing there are the photographic archives, the drawing collection and the registration area.

The objective of the rehabilitation has been to recreate the buildings' original character and emphasize the structure in a dialogue with new elements and spatial constellations. This has been done by demolishing the awkward additions from more recent times and partly re-establishing, adapting and improving the original spatial sizes. Technical installations have been tailor-made and fitted to existing constructions in order to accommodate the antiquarian and architectonic relations in a satisfactory way.

The idea behind the Pavilion was to create an introverted situation, yet one where daylight, sky and surrounding vegetation are nevertheless important elements for the experience of the space. The Pavilion's ground plan is a square with four powerful columns that support a vaguely shell-formed ceiling in light concrete. The glass facades suggest a mere thin layer between outside and in. The Pavilion is encompassed by exterior concrete walls that interplay with neighboring Akershus festning (Akershus Fortress), extend the experience of the space and function as a calm backgrouwnd for the exhibitions.

In the Grosch building the flooring is of brick and polished light concrete on the first floor and white pine in long, knot-free lengths on the second floor. The walls are white-washed and painted with silicate, emulsion and linseed oil-based paints. Original ceilings on the second floor have been re-plastered, and moldings repaired and reconstructed. New fixtures and structural elements are primarily of oak, glass, marble and steel.

The Pavilion has been cast in situ in light concrete. The facade is of glass and the floors of oak.

Text by Martin Dietrichson

Opposite: Interior view of the new exhibition pavilion.

对页：新展览馆内景。

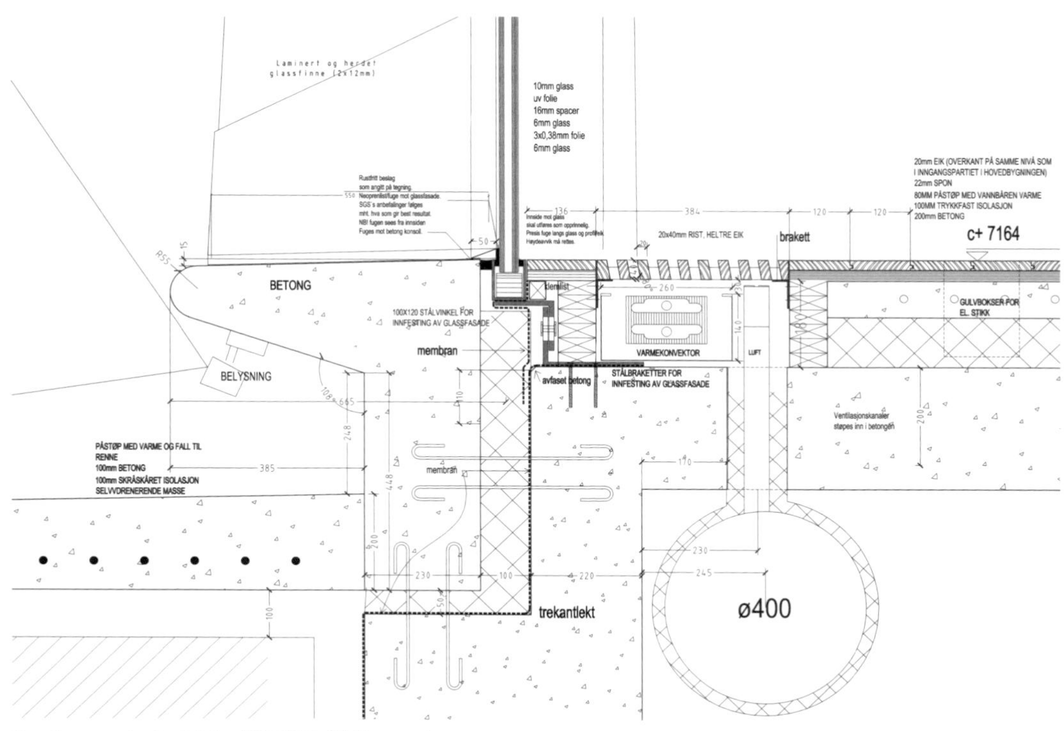

Detail section (scale: 1/15)／剖面详图（比例：1/15）

挪威中央银行在奥斯陆的第一家分行位于银行街3号，由克里斯蒂安·格罗奇设计。它于1830年建成，并成为首座里程碑式的帝国风格建筑。1910年，挪威国家档案馆接管这片土地并增建了4层高的档案馆。自1990年起这座建筑就一直处于空置状态，由于年久失修并经历了几次不成功的改建，受到了一些损伤，但它的地位还是得到了挪威历史遗产局的保护。

2001年挪威国家建筑博物馆向文化部提交了一份将馆址迁移到银行街3号的申请。这项申请得到了媒体的大力支持。同年4月，政府批准了这项申请。在2003年挪威国家建筑博物馆成为挪威国家美术馆建筑设计馆的一部分。此后，这个项目得到了更好的推进，并于2008年1月竣工。

斯维勒·费恩同时担任旧建筑翻修和西南朝向的新展馆的建筑师。新展馆的出资方为詹斯·乌尔特维特·莫，而格罗奇馆和档案馆则由文化部出面融资修建。

博物馆的主入口位于国王街上的建筑原正面入口部分。访客们自入口被引导至地上一层，那里配置了供访客使用的全部功能。在格罗奇馆中设有一个主厅，主厅中配有接待处、书店、咖啡馆、展览空间、卫生间和衣物存放处。访客可经过这个主厅前往特展的展馆，接着便可到达用于常设展的档案馆。朝南设有一个通向公园的出口，夏天可供人们户外就餐。无障碍入口、员工入口以及后勤入口位于下城堡街，而管理中心、图书馆和集会场所则位于格罗奇馆二层。保管照片档案、绘画和图纸集的地方以及登记区则设在档案馆上面的两层。

翻修的主要目标是重构建筑原有的特色，并在新元素、新空间组织之间的对话中强调新结构。具体的做法包括拆除近年来不理想的加建部分，以及对原有空间尺度进行重构、调整和优化。专门定制的技术设备被安装在建筑之中，使原有布局与建筑关系都得到满足的同时，确保空间舒适性。

展览馆背后所蕴含的立意是，在创造一个内向空间的同时，让外部的阳光、天空和周边植物也参与进来，成为提升空间体验的重要元素。展览馆的平面呈正方形，内设4根坚固的柱子支撑着屋顶，屋顶由轻质混凝土制成，略微呈壳形。玻璃立面在室内外形成一道纤薄的隔断。展览馆被外部的混凝土墙包围，与邻近的阿克斯胡斯城堡相互呼应，使空间体验与建筑机能得到了延伸，为在这里展出的展品提供了沉静氛围。

格罗奇馆内一层的地面铺有砖块和打磨过的轻质混凝土，二层则铺设细长且没有木节的松木地板。墙面刷成白色，并施以硅酸盐乳化剂和亚麻油涂料。二层原来的天花板被重新涂刷，线脚也得以修复和重新涂装。新设备和结构元素主要由橡木、玻璃、大理石和钢铁组成。

展览馆在现场以轻质混凝土建成，立面为玻璃，地板为橡木。

马丁·迪特里克松/文

p. 112: Details of glass fin supporting the glass facade. p. 113: Corridor leading into the new exhibition pavilion from the existing building. Opposite: Interior view of the refurbished old building.

第112页：支撑玻璃幕墙立面的玻璃肋细节。第113页：自原有建筑通向新展示馆的走廊。对页：原有建筑翻修后的内景。

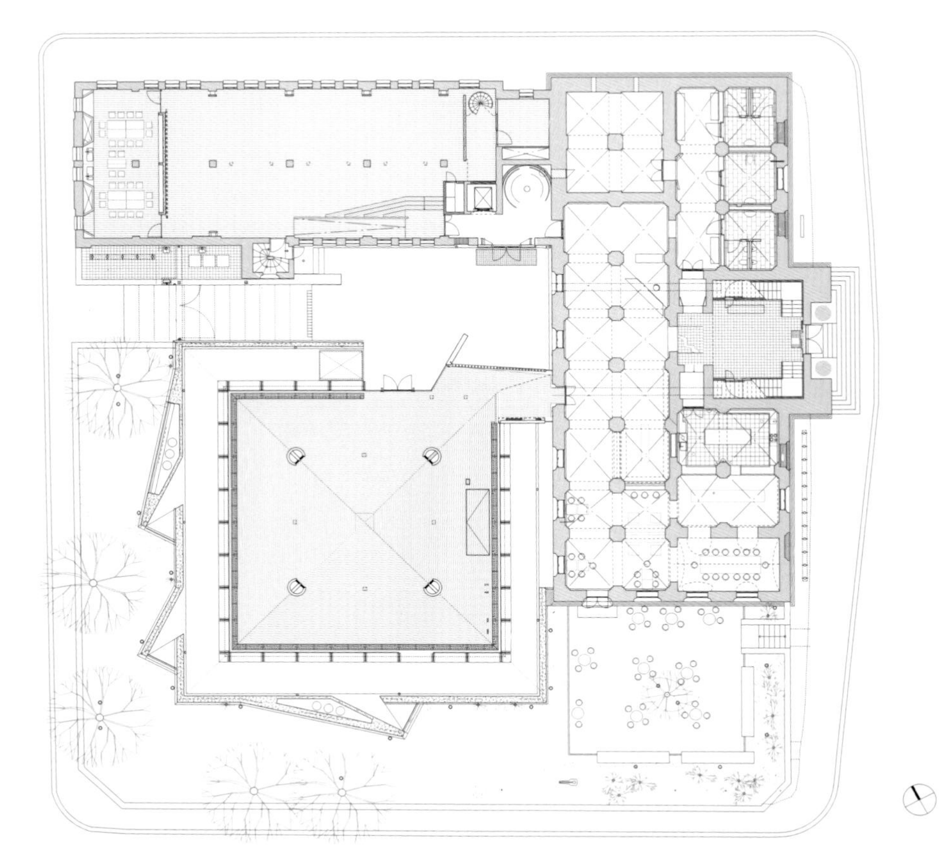

Ground floor plan (scale: 1/500)／一层平面图（比例：1/500）

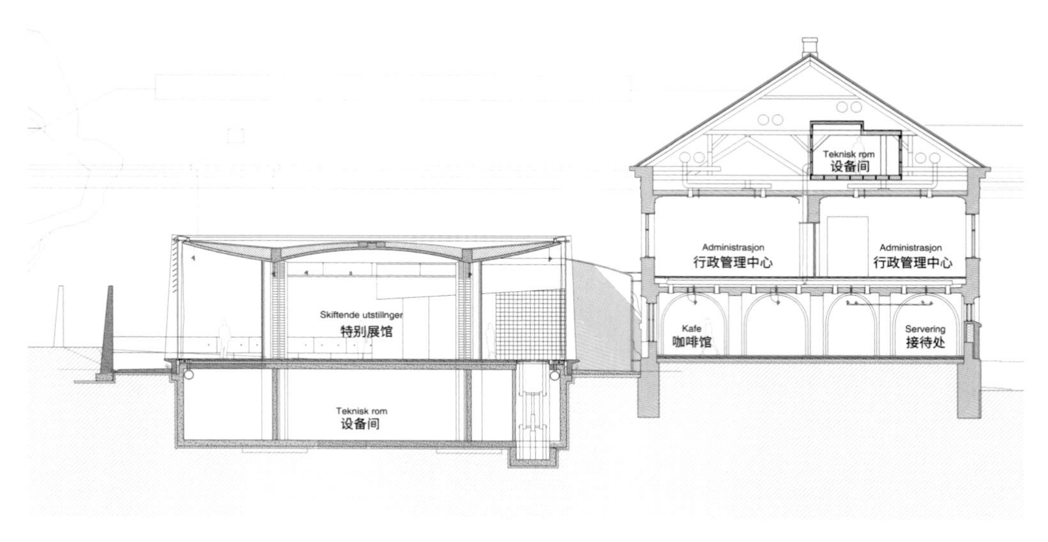

Section (scale: 1/350)／剖面图（比例：1/350）

This page: View from the gap of the pavilion's encircling wall, with a glimpse of the castle.

本页：从环绕展览馆的围墙间隙中可以看到城堡一角。

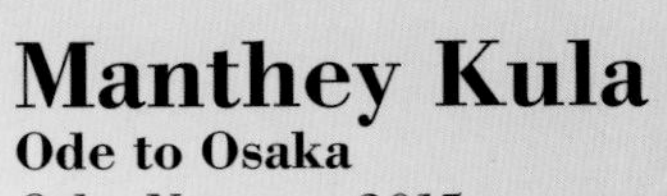

Manthey Kula

Ode to Osaka

Oslo, Norway 2015

曼蒂·库拉

大阪颂

挪威，奥斯陆 2015

pp. 118–119: Interior of the "breathing balloon" pavilion. This page, above: Installation after inflation. This page, below: Installation before inflation.

第118-119页："呼吸气球"内部。本页，上：装置充气后的样子；本页，下：装置充气前的样子。

Architect Sverre Fehn's competition entry for the Scandinavian pavilion at the World Fair in Osaka 1970 is part of the Norwegian National Museum's Architecture collection . Fehn's proposal of a breathing structure where images of Scandinavian nature were to be projected on expanding and contracting walls did not win the competition and was never realized. The inflated structure is atypical in Fehn's production, but the iconic images of the flexible, moving structure still have their power.

As part of a strategy for activating its collection the museum commissioned architects Manthey Kula to develop a concept for realization of Sverre Fehn's competition entry. The task was very open: What was envisioned was a functioning scale model. However, the idea to develop the project and to have it built inside the museum pavilion – which is Fehn's last built work – was presented quite early and supported by the museum.

The installation on show is a result of a design process where important questions concerning the solution of the built piece and its relationship to the initial competition entry had to be addressed and sorted out: The questions concerned technical issues, matters of form and material, geometry, size and siting, and eventually that of exhibition content.

The Ode to Osaka pavilion is not Sverre Fehn's project for the Osaka World fair. It is a contemporary installation based on and honoring his idea of a breathing space. It is a structure consisting of an airlock building and an inflated moving space. All details are developed for the installation to be dismounted and re-erected. There are no objects on show – only space.

Credits and Data

Projec titlet: Ode to Osaka
Client: The National Museum - Architecture
Program: Museum Installation
Location: Bankplassen 3, Oslo, Norway
Design: 2014
Built: 2015
Architect: Manthey Kula
Project team: Per Tamsen and Beate Hølmebakk, responsible architects Magnus Høyem
Structural engineer: Siv. Ing. Finn-Erik Nilsen
Textile and Pneumatics: Luft & Laune, Zürich
Woodwork: Jens Posberg Mortensen
Area: 116 m²
Cost: 1.9 M Nkr

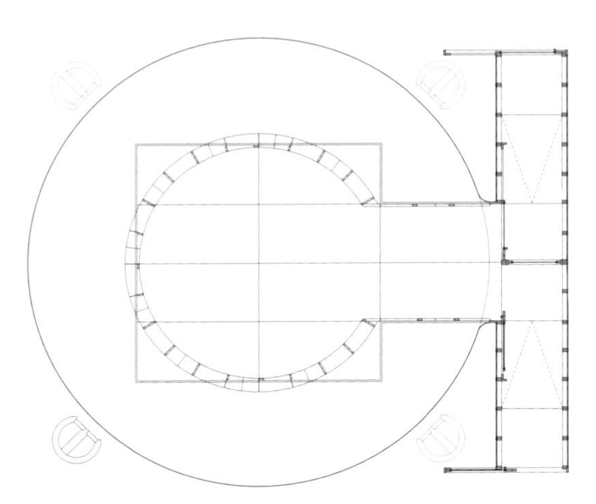

Plan (scale: 1/200) ／平面图（比例：1/200）

这是建筑师斯维勒·费恩为1970年大阪世博会的斯堪的纳维亚馆竞标所设计的方案，现已成为挪威国家博物馆建筑展览的一部分。费恩的方案是一个仿佛能呼吸的膨胀结构，代表了斯堪的纳维亚地区的自然环境。但这个方案并未赢得竞标，也未能实现。膨胀的建筑结构在费恩的作品中十分少见，但它标志性的自由形态、可活动的结构至今仍拥有着超前的力量。

为了给建筑展品注入新的活力，挪威国家博物馆委托曼蒂·库拉建筑事务所重现斯维勒·费恩的竞标方案。项目内容看似非常开放，即将一个概念模型在既定比例的基础上变成现实的建筑；但是，将费恩最后的建成项目，也就是他当年的竞标作品，在这座博物馆的展厅中重现是博物馆一直以来的愿景。

本项目在设计过程中不断解决实际落地时遇到的重要问题，同时也在思考它与真正建成的世博会展馆间的关系。问题包括技术因素、形式和材料、几何学、尺寸和定位，以及最后与展品之间的关系。建成后，这个项目本身就是对以上问题与思考的解答。

大阪颂展馆并非原本斯维勒·费恩为大阪世博会设计的方案，而是一个向费恩的“呼吸之肺”理念致敬的现代构造。它由一个空气密封构造和一个可缩放的空间构成，也就是说，这是一个再膨胀的项目计划。所有部件都被设计成可拆卸、再组装的形式。这里没有任何物品展出，只有纯粹的空间。

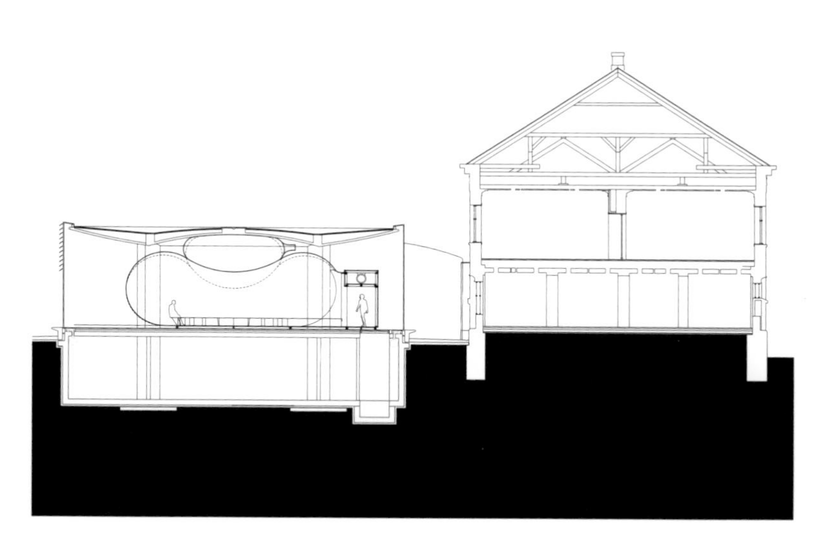

Section (scale: 1/400) ／剖面图（比例：1/400）

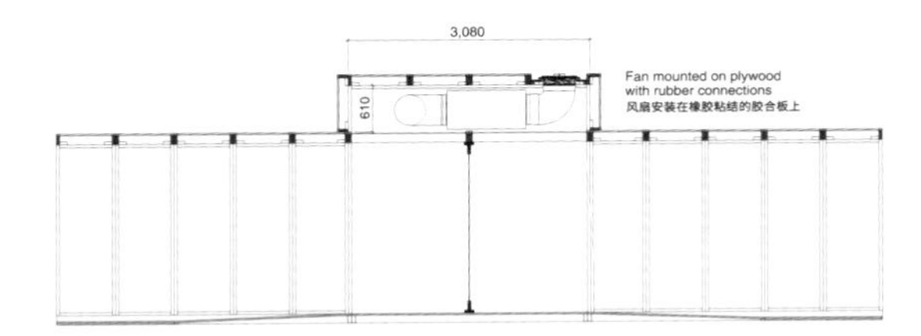

Fan room long section (scale: 1/150) ／风机室纵剖面图（比例：1/150）

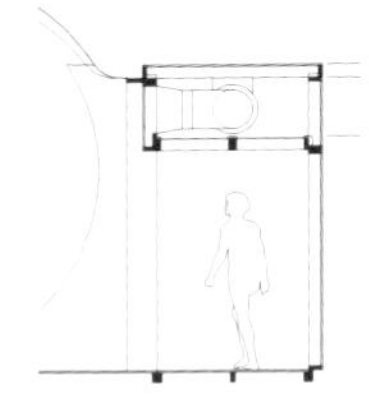

Fan room short section ／风机室横剖面图

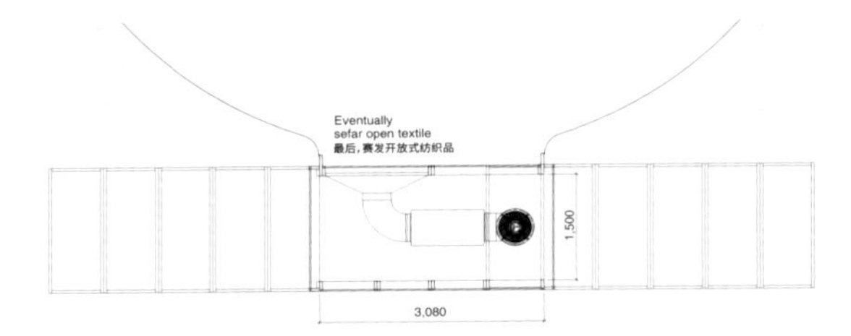

Fan room plan (scale: 1/150) ／风机室平面图（比例：1/150）

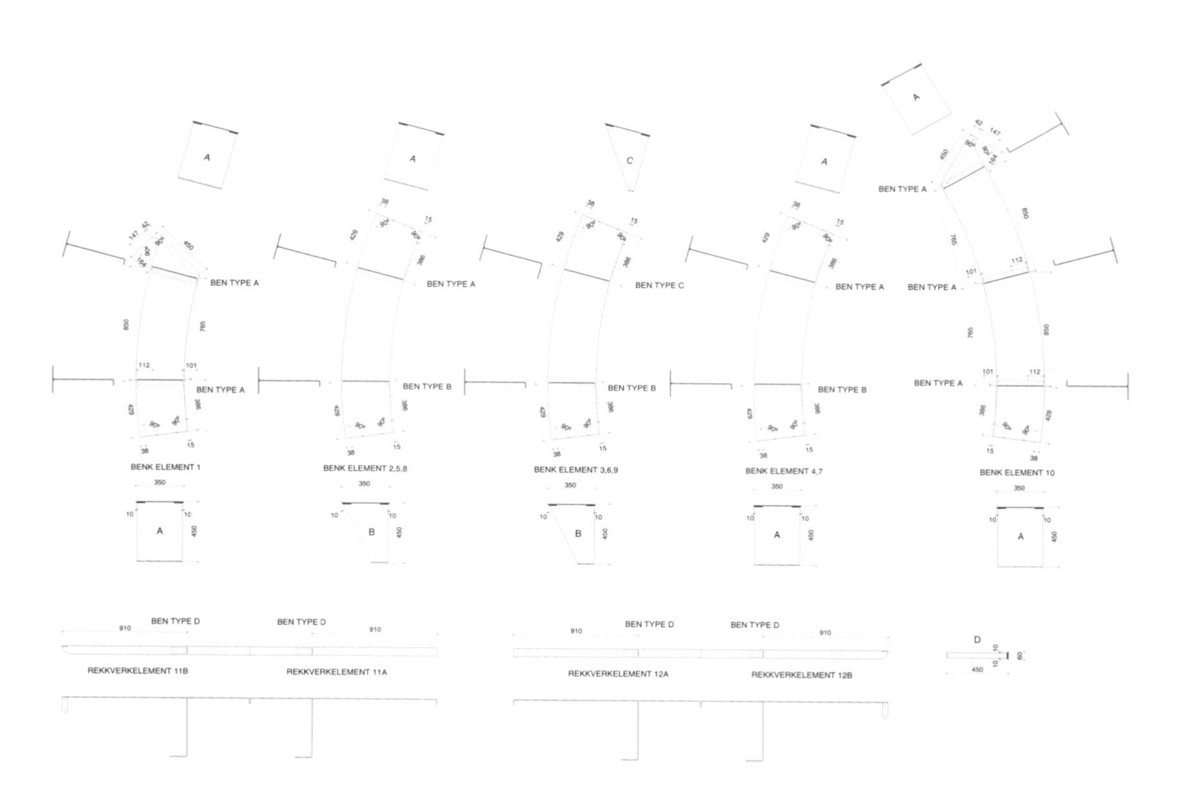

Detail of bench (scale: 1/55) ／长椅详图（比例：1/55）

p.123: The entrance of the "breathing balloon" pavilion. Opposite: Details of the bench. This page: No objects on show – only space. All photos on pp. 118–125 courtesy of the Architect.

第123页："呼吸气球"的入口。对页：长椅的细部构造。本页：没有任何物品展出，只有纯粹的空间。

Manthey Kula
Skreda Roadside Rest Area
Leknes, Norway 2018

曼蒂·库拉
斯克里达路边休息区
挪威，莱克内斯　2018

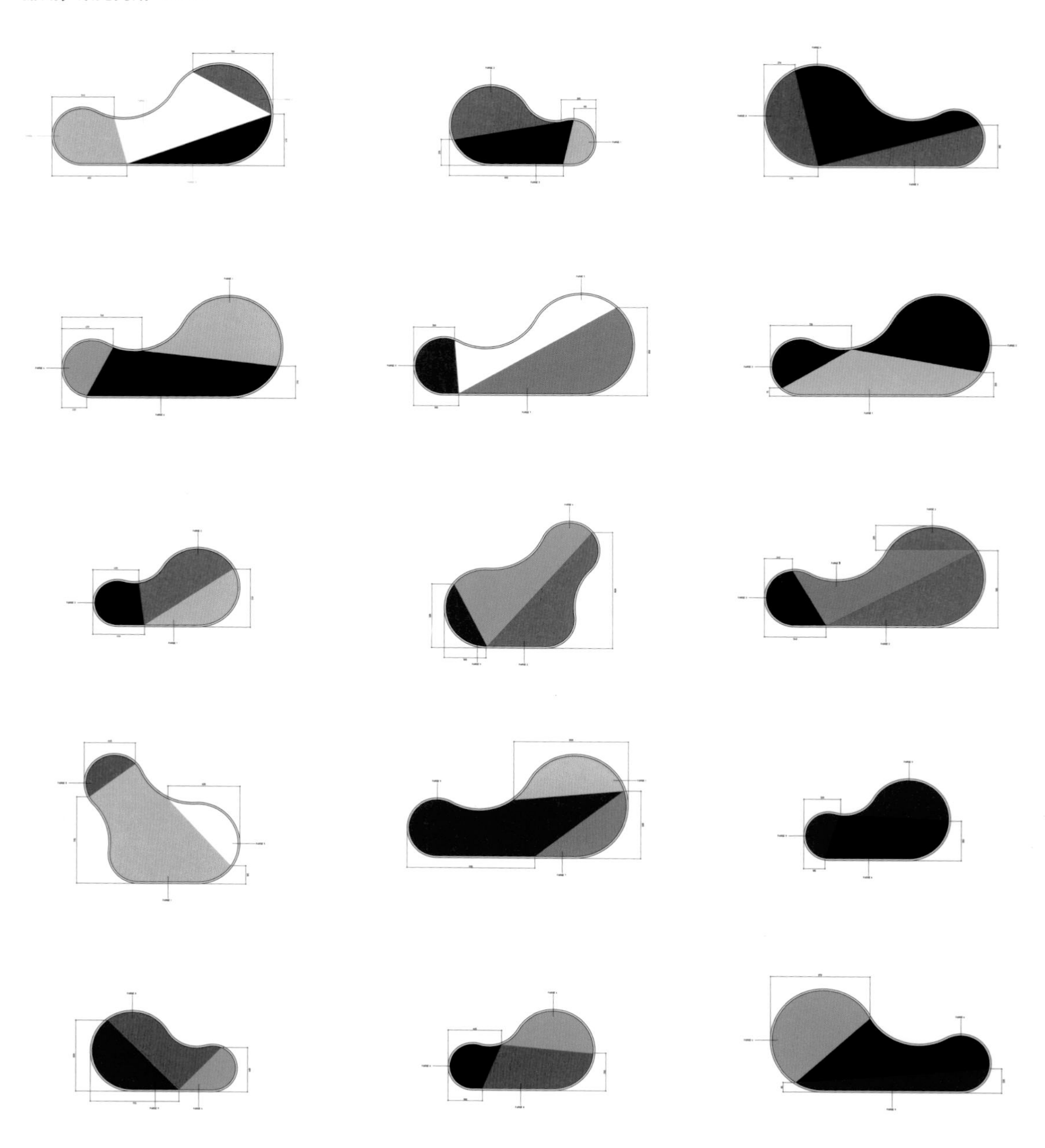

Furniture plan (scale: 1/60) ／家具平面图（比例：1/60）

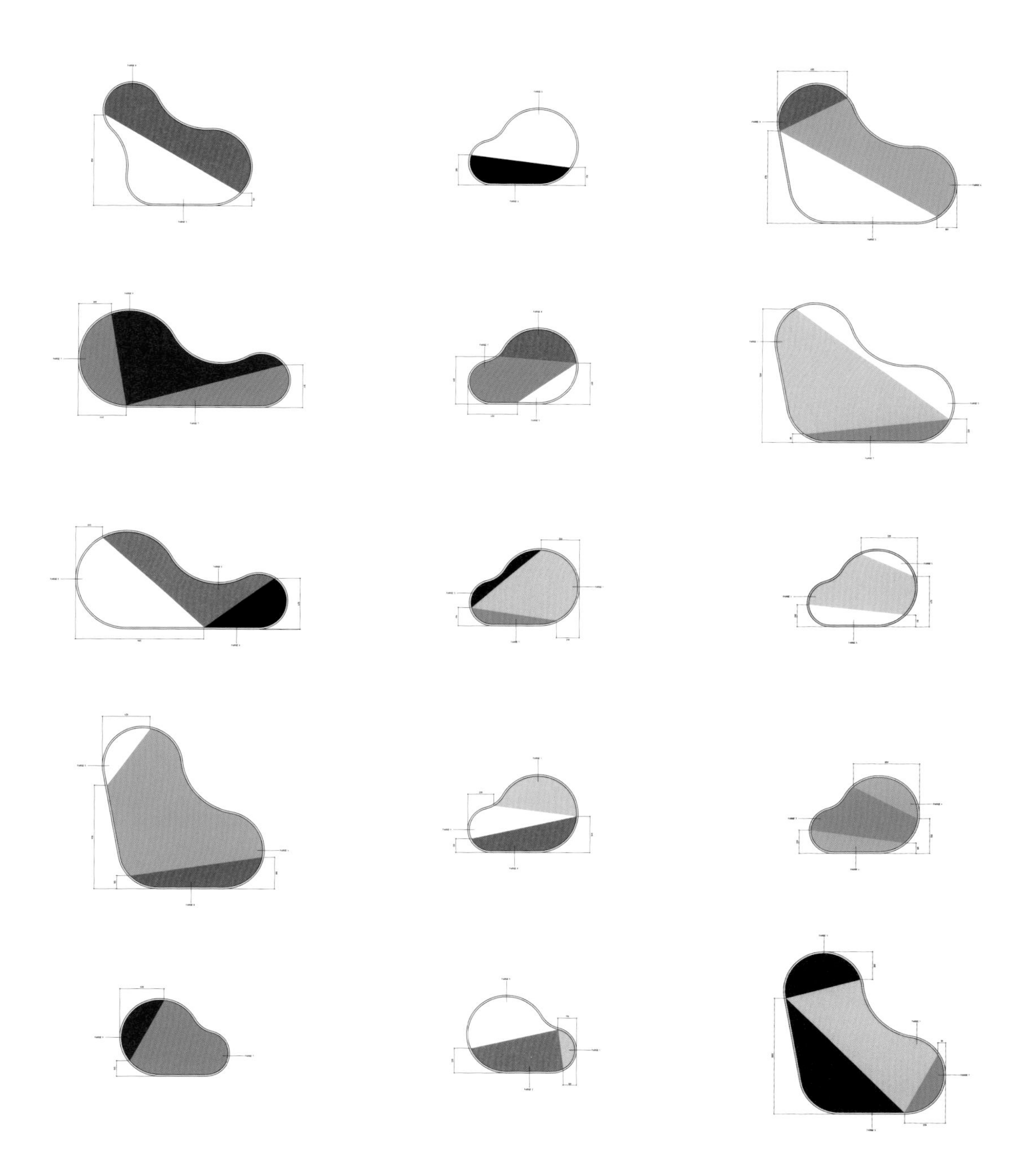

This page and opposite: Curved steel sheets are painted with triangular patterns in solid colours. All photos on pp. 126–137 courtesy of the Architect.

本页和对页：弧边造型的钢板表面涂有三角形的纯色图案。

The project is a complete transformation of an existing rest stop which was established in the 1980s. The old rest area was organized in and between an artificial terrain of nature-like mounds and boulders which obscured the view to the Lofoten seascape, the place was dominated by an oversized traffic area.

The new project aims to establish an inviting floor where travelers can find pleasure in searching for seating and from where they can enjoy the beautiful view across the water. The platform is subordinate to the landscape, yet its convex geometry contributes to place the visitors in an exposed but secure position in the center of the large open space.

The new project consists of the platform which is constructed of steel sheets span between concrete foundations. These foundations form floors for the furniture, ramps and stair. The furniture consists of benches, tables and railings, all structurally interdependent, all made from bent standard flat steel bolted or welded together to form stable elements. There are thirty different surfaces to sit on and by. Each piece has a different pattern in solid colours painted on the steel surface, covered with a cast of clear polyurethane. This material is warmer to touch than steel and highly reflective; mirroring the ever-changing sky. The painted pattern indicates the construction of the furniture. Each piece is only directly connected to two legs, the third supporting leg is placed outside the piece itself, as part of the adjoining handrail. In this way, a triangular pattern is established. The slope between the road and the rest area, with its strict geometric form, is high enough to reduce noise from the passing cars and low enough to not completely obscure the view. The slope is planted with Salix myrsinifolia ssp. Borealis, a small tree indigenous to the northern areas. The traffic area is reduced in comparison to the initial situation. It is the curvature of this area that has contributed to the main geometry of the project.

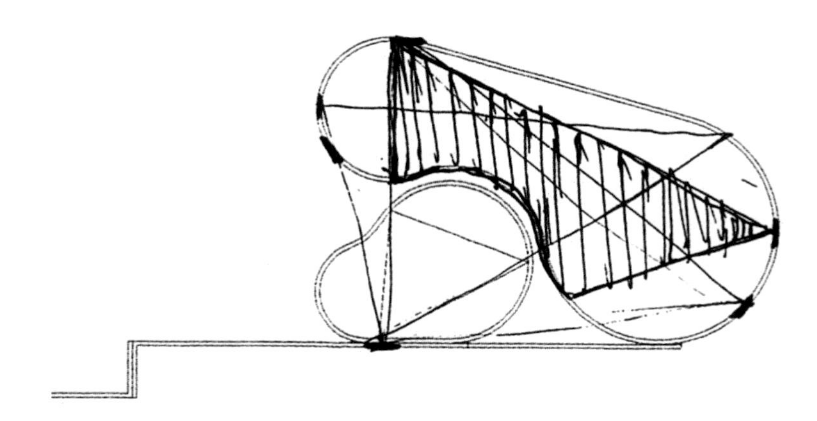

Opposite: Curved steel sheets are painted with triangular patterns in solid colours.

对页：弧边造型的钢板表面涂有三角形的纯色图案。

Credits and Data

Project title: Skreda Roadside Rest Area
Client: National Tourist Routes in Norway
Program: Public Infrastructure
Location: Flesveien, 8370 Leknes, Norway
Design: 2014–2016
Completion: 2018
Architect: Manthey Kula
Team: Beate Hølmebakk and Per Tamsen, responsible architects with Magnus Høyem, Nina Fjose and Frida Johansen
Structural engineer: Siv. Ing. Finn-Erik Nilsen, Dr. techn. Kristoffer Apeland, Snorre Larsen
Area: 6,000 m²
Cost: Ca. 10 M Nkr

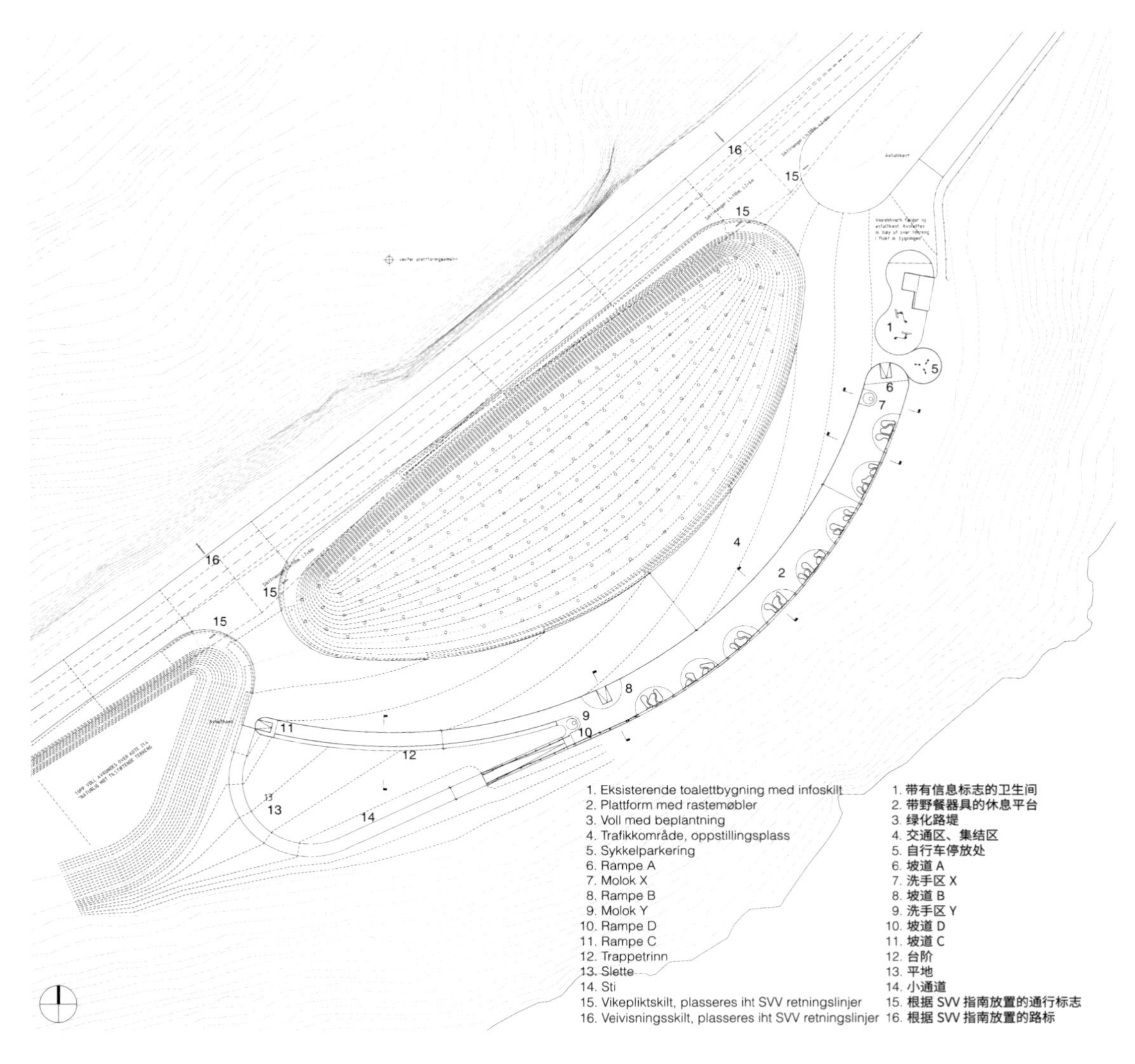

Site plan (scale: 1/1,000) ／总平面图（比例：1/1,000）

pp. 132–133: Each set of furniture sits on a concrete foundation, facing a view across the water. Opposite, both images: View during construction. p. 136: Furniture unit consists of benches, tables and railings.

第132-133页：每组家具都设置在一个混凝土地基上，面朝水域对岸的景色。对页，上图和下图：建设过程中的景象。第136页：由长椅，桌子和扶手构成的家具组合。

本项目是对一个建于20世纪80年代的休息区的彻底改造。休息区位于一片模仿自然山丘和岩石的人工地形中。罗弗敦群岛的海景被周围的地形遮挡，而这片休息区也被过大的交通区域所支配。

新项目希望建立一个具有吸引力的场所，游客在此落座休息时可以感受到乐趣，并能欣赏对岸的美景。休息平台是景观的一部分，而它凹凸出的几何造型为游客提供了一处能安全停留的巨大开放的空间场所。

新项目由架设在混凝土地基上的钢板平台组合而成。这些地基也自然地成为家具、坡道和台阶的平台。家具包括结构分离、相互依存的长椅、桌子和扶手，它们均由标注尺寸的钢材弯曲、再用螺栓或焊接固定而成，形成稳定的部件。一共有30种不同的桌椅表面，每一块表面都由钢板涂刷成不同图案，并在纯色的表面覆盖了一层透明的聚氨酯材料。这种材料的触感比钢铁更温暖且表面的反射性很强，总是能反射出千变万化的天空。涂刷的图案象征着家具的构造。每片钢板都只与两个支撑直接相连，而第三个支撑则被设置在钢板外，成为相邻扶手的一部分，这样一个三角形的布局便形成了。位于道路和休息区之间的坡道经过了几何学的精密计算，它的高度足以减少来往车辆的噪声，又不至于完全遮挡住视线。坡道上种有青蓝柳，这是一种生长在北方的矮小树木。交通占用的面积小于改造前。正是区域本身的曲折形态造就了该项目的主要几何形态。

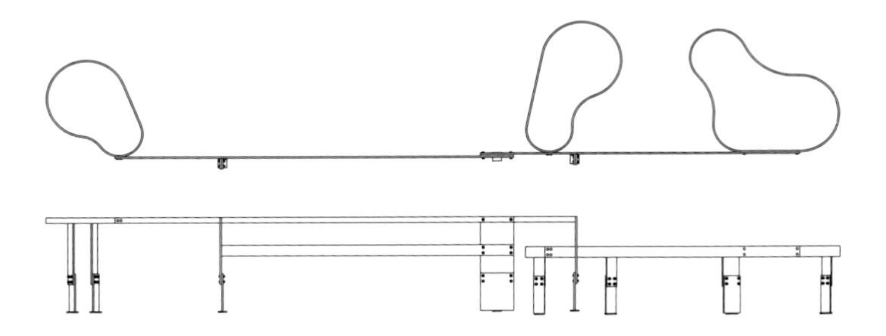

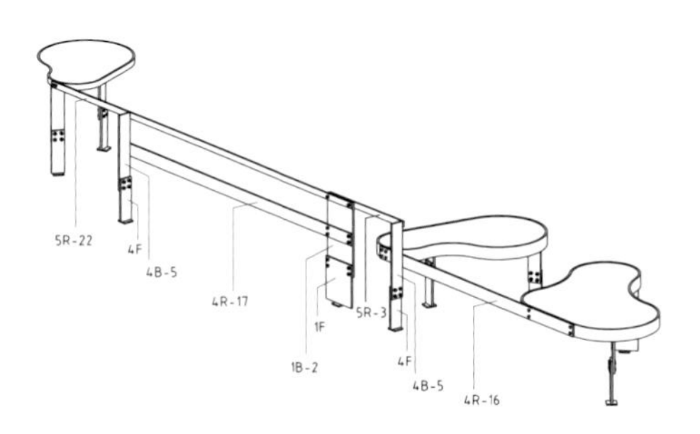

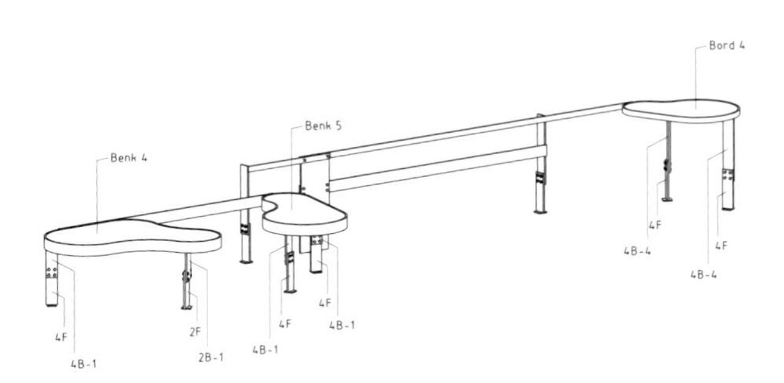

Detail of a furniture unit (scale: 1/80) ／家具单元详图（比例：1/80）

Manthey Kula

Forvik Ferry Port

Vevelstad, Norway 2015

曼蒂·库拉

福维克轮渡港口

挪威，韦韦尔斯塔 2015

The Place

Traveling by car along Norway's coast the journey is often brought to a halt by lines of vehicles waiting for a ferry to cross one of the many fjords. At Forvik on the northwestern coast the ferry formerly landed on a dock of an 18th century tavern. The old dock could no longer support the terminal function; hence it was decided to build a new port on the other side of a small bay, leaving the tavern without direct contact with the traveling crowds.

Our task was therefore – in addition to designing the landscape intervention and a small service building for the travelers – to establish a visual connection between the new terminal and the old tavern.

For the landscape part of the project our focus was to narrow the traffic area and to give the intervention a distinct, but subordinate form. This is achieved by situating the service building at the very end of the traffic area, in direct connection to the new standard dock and by eliminating ditches and retaining walls. Additionally, careful planning of the cutting of bedrock and the forming of new ground and surface drainage helped reduce the intervention in the vulnerable coastal terrain.

The Building

The service building is designed to be both a visual gateway and a shelter. Its dominating structural element, the inverted steel vault, evokes the idea of a large sea creature or of the hull of a ship. It spans between, and cantilever past, the concrete gable walls. The 10 mm galvanized steel roof stabilizes slender wooden frames below. The frames are clad with insulated glass creating freestanding indoor spaces serving as a kiosk, a waiting room and rest rooms. Open spaces between the glass volumes and under the cantilevering steel roof give outdoor shelter for travelers. The building is completely transparent. The underbelly of the large steel roof can be seen from every room and so can every structural element. During the night and the long dark winter months the inverted vault is lit to show the way to the travelers.

Except for the concrete foundations, stairs and gable walls the building is made from prefabricated parts that are bolted together: rolled and/or folded steel elements for the roof, welded bits for bracing and railing, glue-lam timber frames of different shapes, grates of many sizes, panes of single or double glass, pre-cut sheets of Finnish plywood, doors and modest furniture. The main contractor who assembled the parts on site was a specialist in road construction and had never built before.

Credits and Data

Project title: Forvik Ferry Port
Client: Norwegian Public Roads Administration
Program: Public Infrastructure
Location: 8976 Vevelstad, Norway
Design: 2012–2013
Built: 2015
Architect: Manthey Kula
Team: Per Tamsen and Beate Hølmebakk, responsible architects with Jonas Larsen, Frida Johansen, Magnus Høyem and Nina Fjose
Structural engineer: Siv. Ing. Finn-Erik Nilsen, Dr. techn. Kristoffer Apeland, Snorre Larsen
Construction: Norwegian Rock Group
Area: Building 120 m², Landscape 19,000 m²
Cost: 45 M Nkr

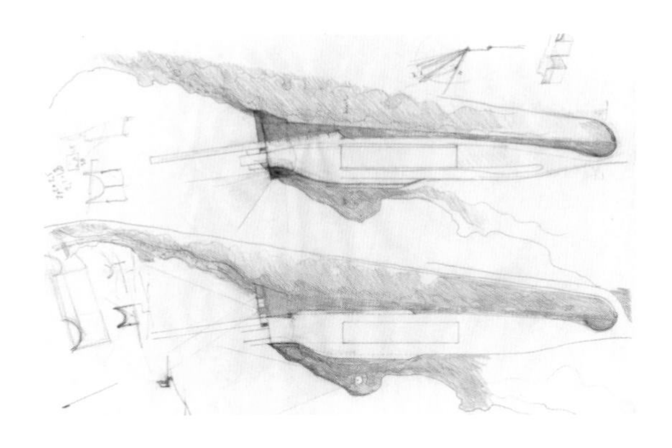

场地

当人们驾车沿着挪威海岸线旅行时，由于需要排队等候轮渡穿越海峡，行程经常被中断。位于西北海岸福维克的轮渡从前一直停泊在一座18世纪旅馆的码头上。老旧的码头无法继续承担轮渡站的功能，因此当地决定在一个小海湾的另一侧修建一个新港口，避免旅馆与游览人群发生直接接触。

因此本项目的目标是为游客设计一座小的服务中心以及配套的景观规划，更重要的是在新轮渡站和旧旅馆之间建立起视觉联系。

在项目的景观规划方面，我们的关注点在于交通区域的缩窄以及让这次改造在形式上明确而不喧宾夺主。具体的措施包括将服务中心移到交通区域的尽头，使它直接与新建的标准码头相连；清除沟渠和挡土墙。此外，对岩床的切割、新地表的建造以及排水处理的周密规划，也能够减弱对这片脆弱海岸区域的威胁。

建筑

服务中心被设计成视觉上的出入口，同时又可以作为庇护所。它的主导结构元素——倒置的钢制拱顶，能让人联想到一个巨大的海洋生物或者一条船的船身。拱顶横跨并悬挑出混凝土山墙。10毫米厚的镀锌钢屋顶对下面纤细的木结构起到了稳定作用。框架上覆盖着隔热玻璃，形成了独立的室内空间，用作售货亭、等候室和休息室。玻璃体量和悬空钢铁屋顶之间的开放空间为旅行者提供了室外的庇护所。建筑本身是完全透明的。从每个房间都能看到巨大钢铁屋顶的下方以及所有结构元素。在冬季漫长而漆黑的夜晚中，倒置的拱顶被照明点亮，为旅行者指明道路。

除了混凝土地基、台阶和山墙，建筑的其他部分全部由预制部件连接组合而成，包括卷曲或交叠的屋顶钢部件，支撑结构和栏杆的小焊接部件，不同形状的胶合木构架，各种尺寸的格栅，单层或双层的玻璃窗，提前切好的芬兰胶合板、门和朴素的家具。而装配这些部件的承包商主要从事的是道路修建专业，在此之前没有任何建造建筑的经验。

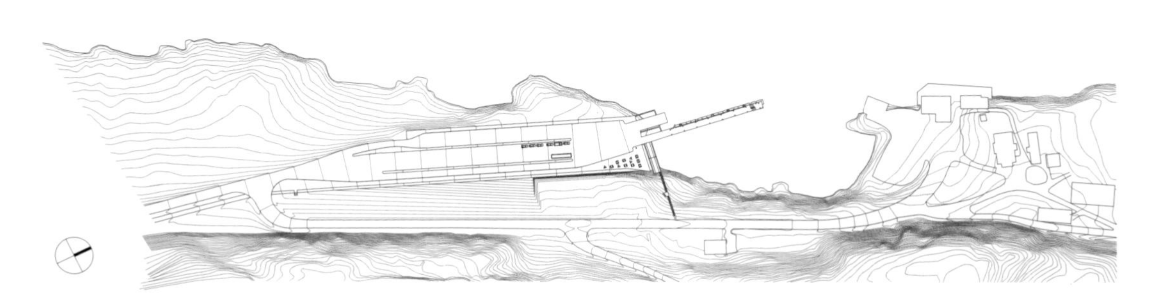

Site plan (scale: 1/5,000)／总平面图（比例：1/5,000）

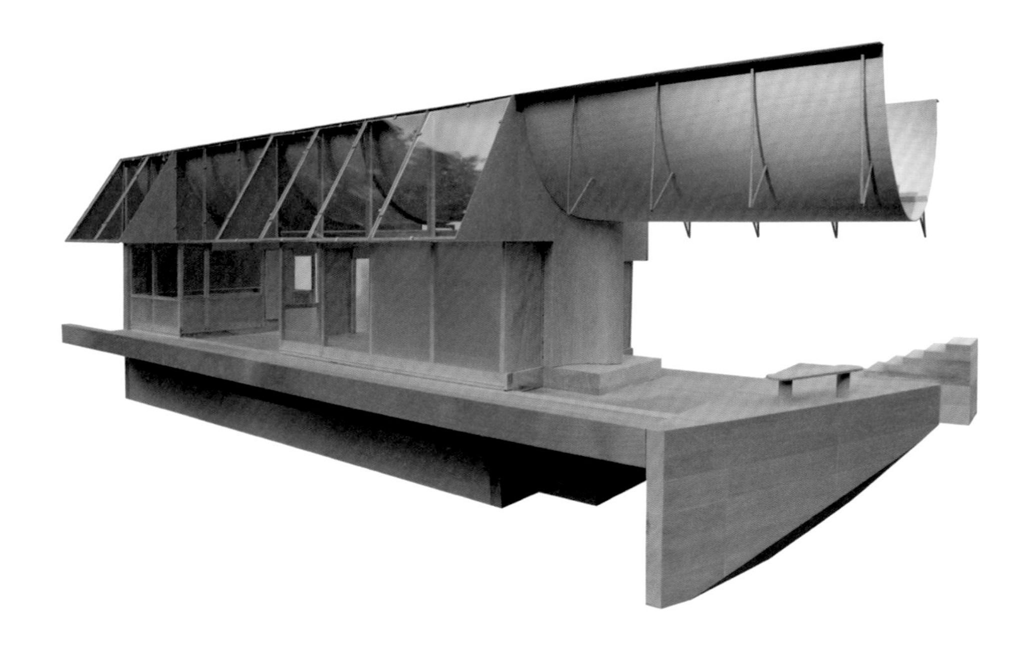

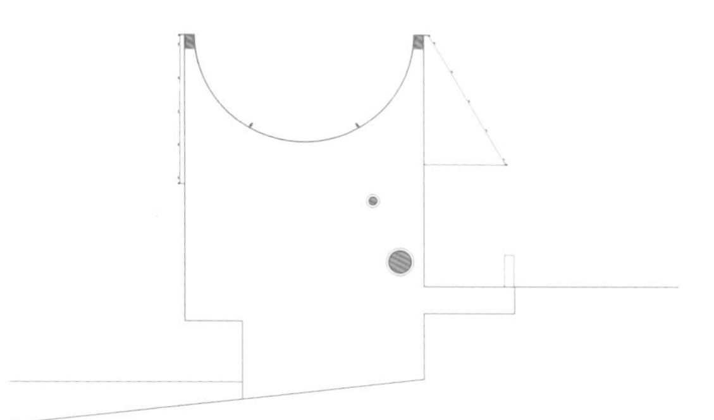

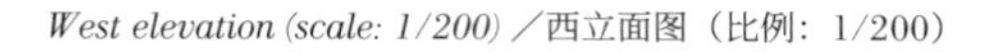

West elevation (scale: 1/200) ／西立面图（比例：1/200）

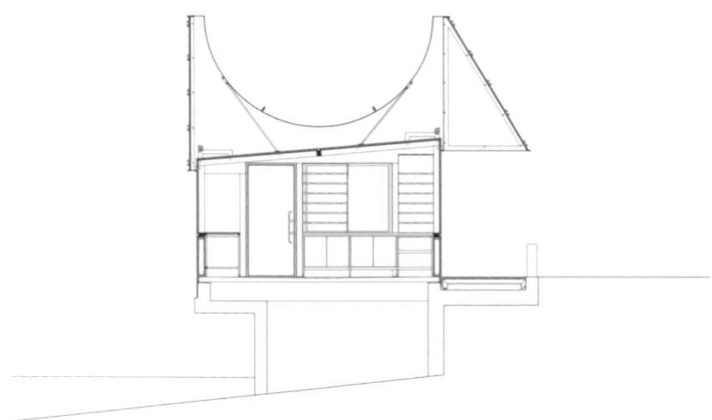

Section A (scale: 1/200) ／ A 剖面图（比例：1/200）

p. 139: Night view looking towards the stairs leading into the entrance. p. 140: Study sketch by the Architect. p. 141, above: Model photo. p. 141, below: View during the construction stage.This page: The inverted steel vault spans between the concrete gable wall, with its sides clad with glass.

第139页：通向入口处楼梯的夜景。第140页：建筑师手绘。第141页，上：模型照片；第141页，下：建造过程中的样子。本页：倒置的钢制拱顶横跨山墙，两面覆盖着玻璃。

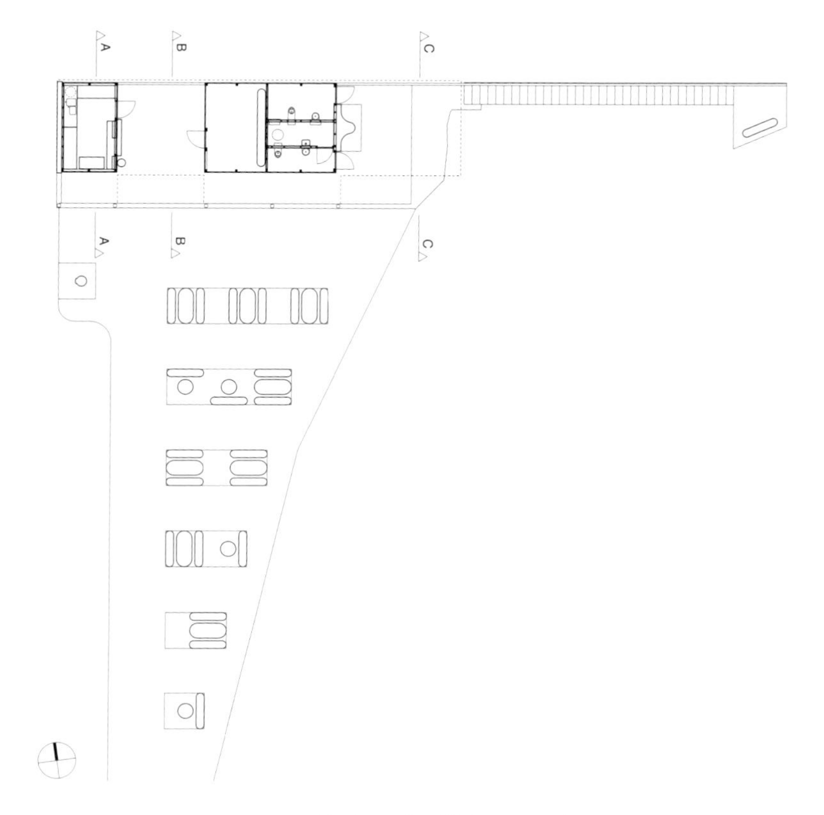

Ground floor plan (scale: 1/1,000) ／地面层平面图（比例：1/1,000）

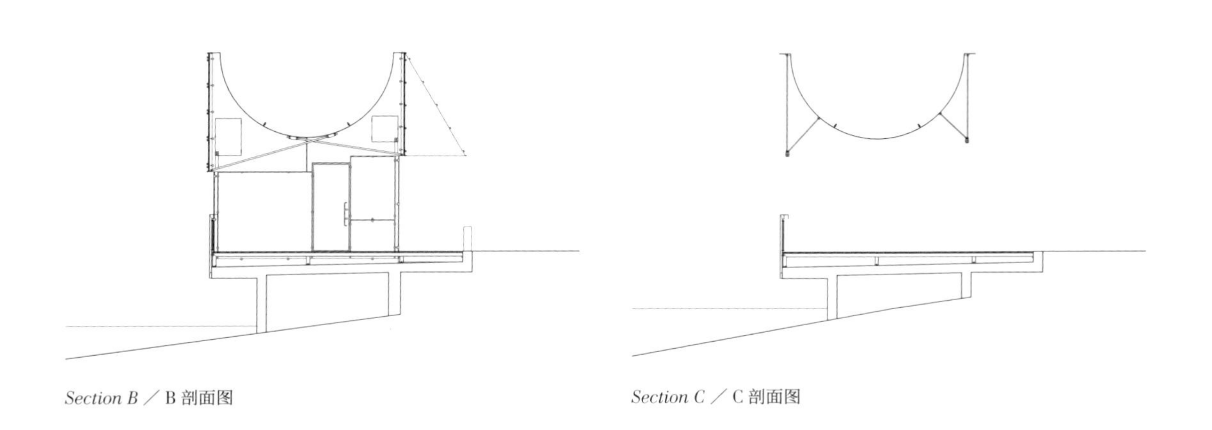

Section B ／ B 剖面图

Section C ／ C 剖面图

p. 145, above: Elevation view of the cantilever inverted steel vault with its structural element exposed. p. 145, below: View looking towards the coast from the corridor. All photos on pp. 138–145 courtesy of the Architect.

第145页，上：悬梁固定的倒置钢屋顶立面，所有结构元素都裸露在外；第145页，下：从走廊处远眺海岸线。

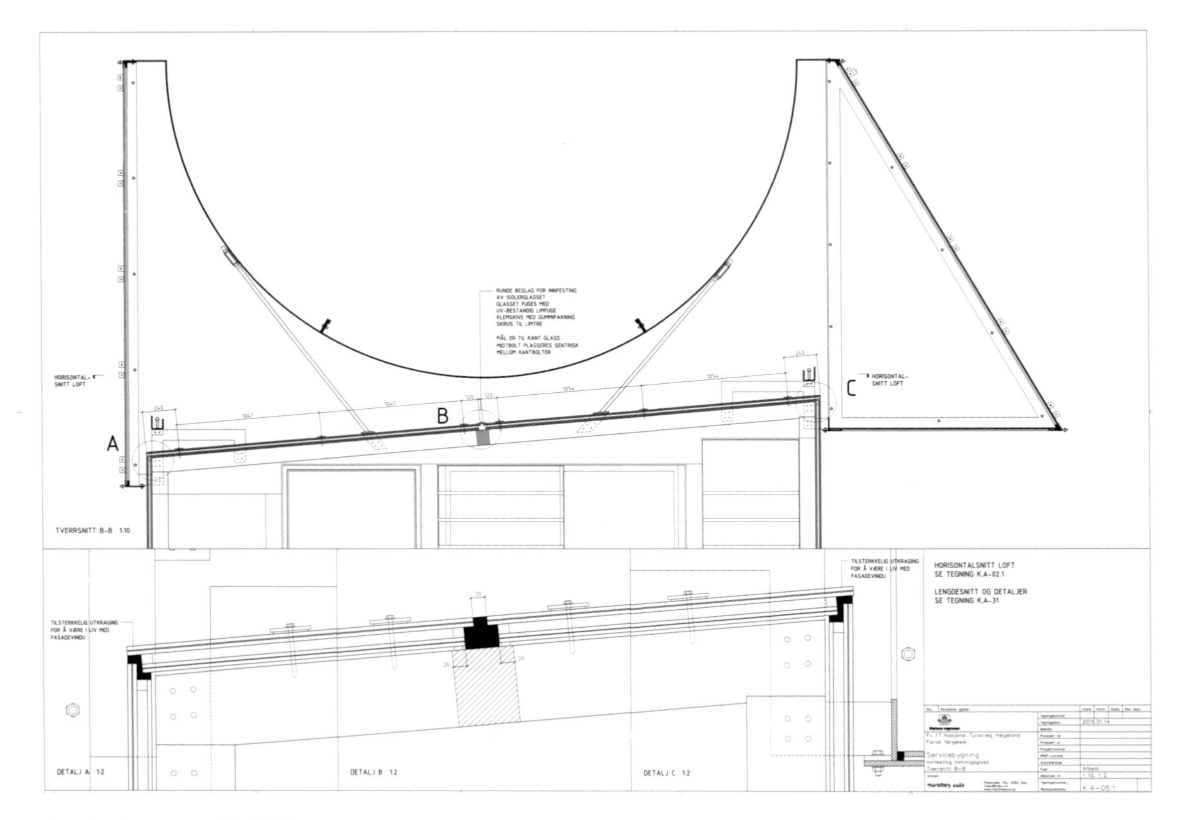

Details of section A ／剖面详图 A

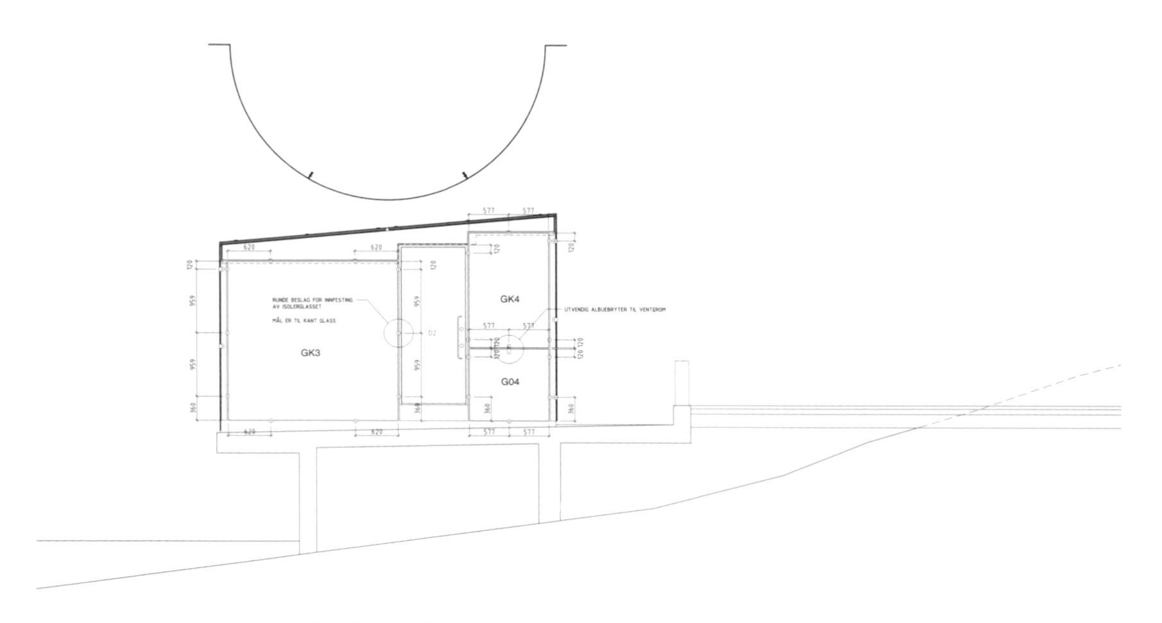

Details of section B (scale: 1/110) ／剖面详图 B（比例：1/110）

Manthey Kula

Stella's Room

Stange, Norway 2016

曼蒂·库拉

斯特拉的房间

挪威，斯唐厄 2016

This project is an added room and a new entrance to the barn building at Sørum farm at Stange, Norway. The farm, situated on a small hill surrounded by fields, dates from the 17th century. The farm as it stands today is the result of a series of architectural interventions by architect Are Vesterlid in collaboration with Sørum's current owner, the sculptor Knut Wold. The result is a rich architectural collage of historic timber tradition and original contemporary work. Stella's room is an addition to this context, executed some years after Vesterlid's passing.

The rooms are built in the un-insulated part of the barn, partly behind an old barn door. A free-standing glass wall on the inside of the existing timber structure constitutes the room's exterior wall, exposing the beautiful order of the old posts and beams. When the large barn door is closed light is brought into the room through a large sculptural skylight.

Credits and Data

Projec title: Stella's Room
Client: Private
Program: Addition inside existing barn garage
Location: Sørum farm, 2335 Stange, Norway
Built: 2016
Architect: Manthey Kula
Project team: Beate Hølmebakk and Per Tamsen
Area: 25 m^2

Opposite: Model of the room. This page: View of a room showing the relationship between existing and new timber. All photos on pp. 146–153 by the Architect.

对页：模型图。本页：一间房间内的样子，展现了新旧木结构之间的关系。

This page, above: Existing old barn door of the entrance, half opened revealing the new timber door. This page, below: Exposed old timber structure.

本页，上：原有仓库入口处的门呈半打开的状态，露出新木门的样子；本页，下：外露的旧的木结构。

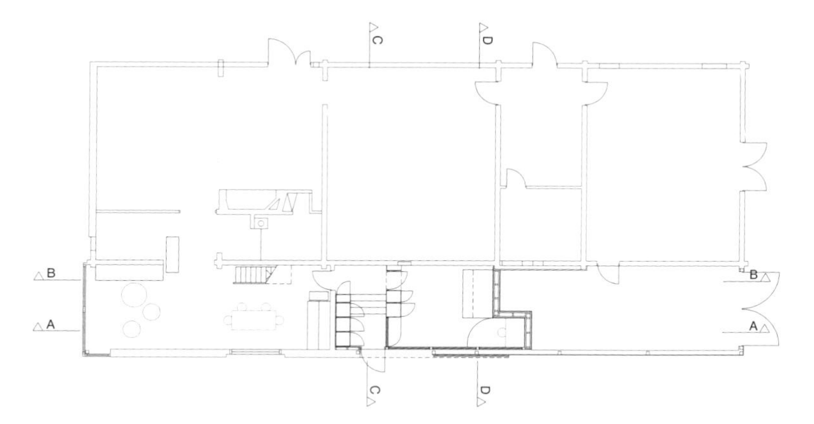

Plan ／平面图

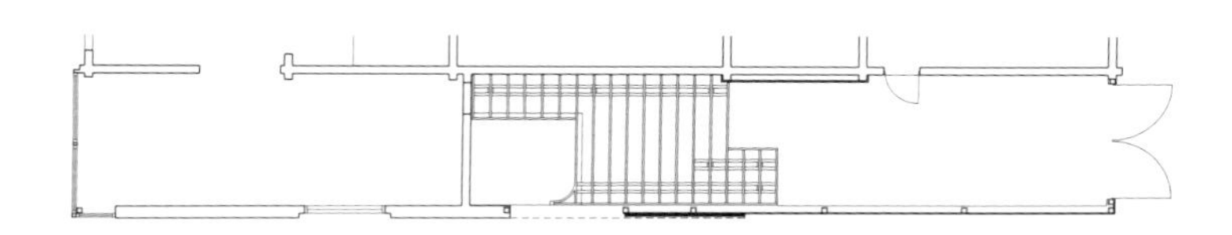

Joists plan ／梁俯视图

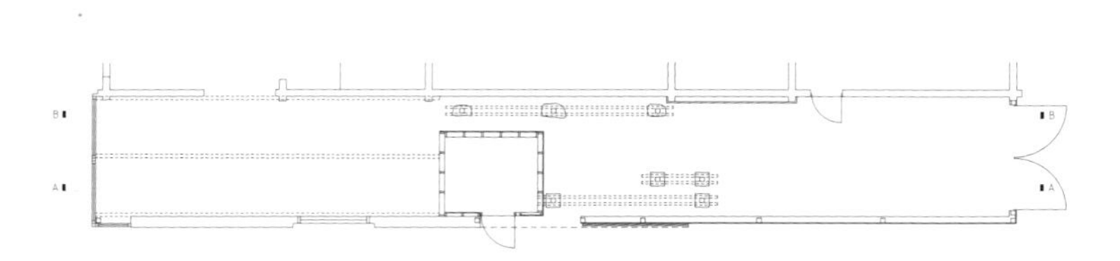

Footing plan ／地基俯视图

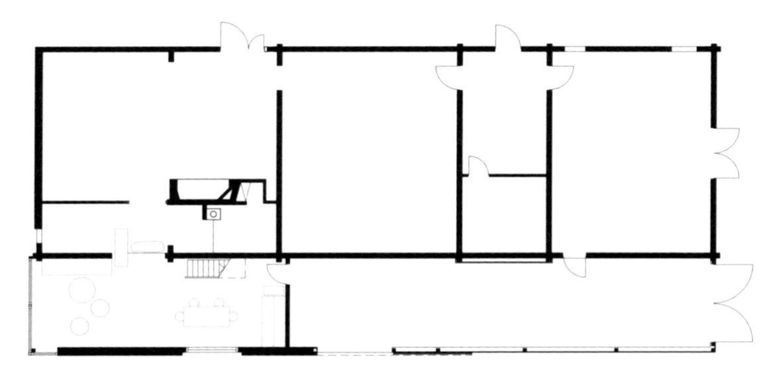

Plan (Existing condition, scale: 1/300) ／平面图（改造前，比例：1/300）

Opposite and this page: Light brought into the room through a sculptural skylight. p.153: Detail of the glass wall.

对页及本页：从一扇具有雕塑感的天窗射进室内的自然光。第153页：玻璃幕墙细部。

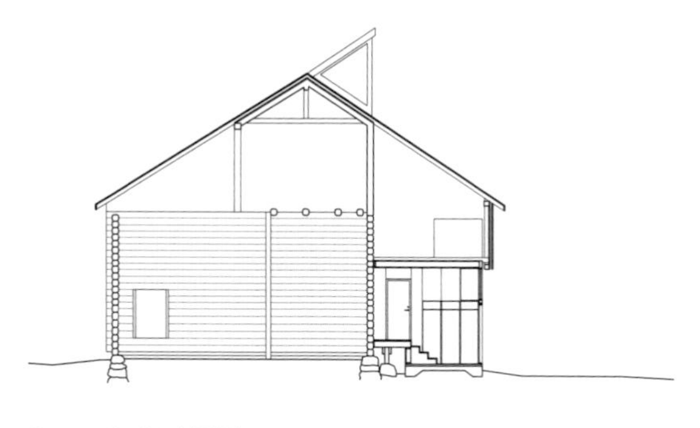

Section C ／ C 剖面图

Section D ／ D 剖面图

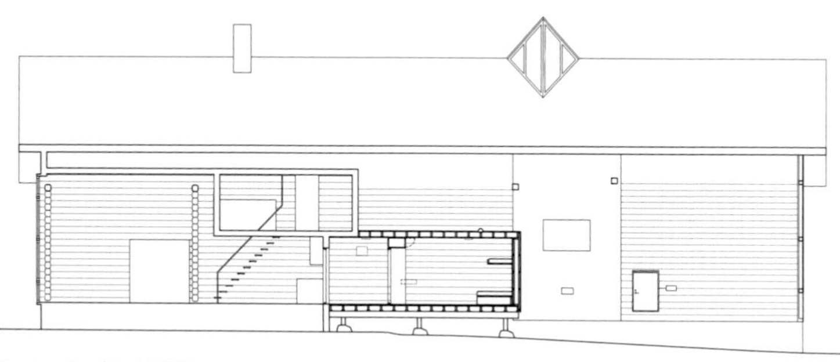

Section B ／ B 剖面图

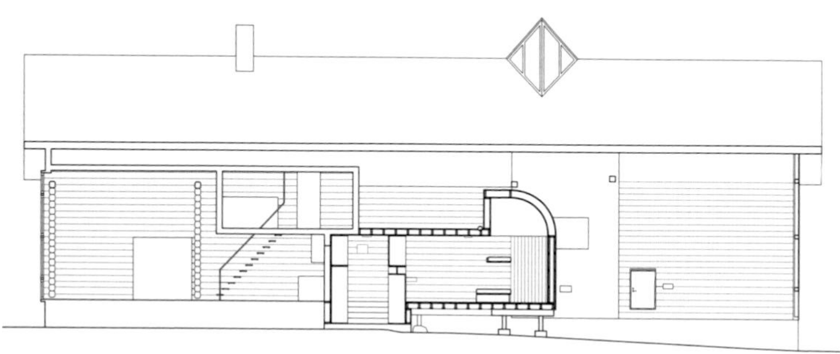

Section A(scale: 1/300) ／ A 剖面图（比例：1/300）

这个项目是一个农场仓库的增建房间及新入口。场地位于挪威斯唐厄市的瑟伦农场，而农场位于一座被农田环绕的小山上，其历史可以追溯到17世纪。建筑师阿尔·维斯特里德和现任农场主人雕塑家昆特·沃尔德的一系列建筑改造活动，造就了农场现今的模样。如今这座农场中同时拥有丰富的历史传统木结构和原创的现代设计。在维斯特里德去世数年后，“斯特拉的房间”这一项目才得以实施，可以视为是对上一位建筑师所营造的环境的新继承。

房间被建造在现有仓库之外，部分隐藏在一扇旧仓库门的背后。一面建造在现有木结构内侧的独立玻璃幕墙构成了房间外墙，露出古朴美丽的柱和梁。当巨大的仓库门关闭时，自然光也可以通过一扇具有雕塑感的巨大天窗进入室内。

Manthey Kula
Tullholmen Pier
Karlstad, Sweden 2017–

曼蒂·库拉
塔尔霍尔曼栈桥
瑞典，卡尔斯塔德　2017-

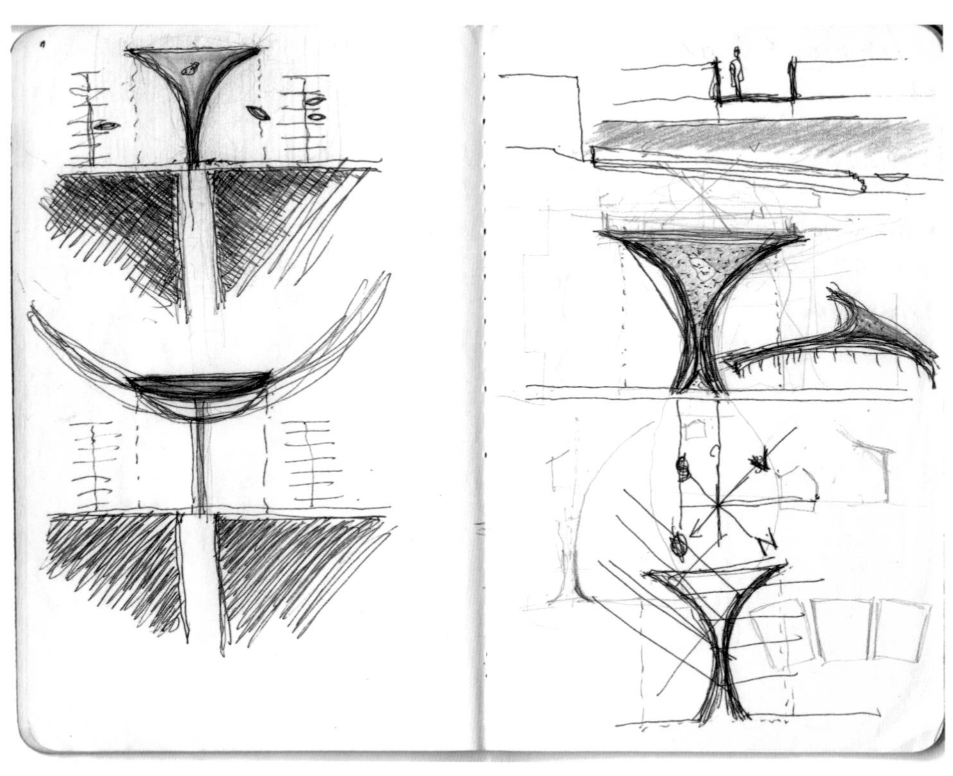

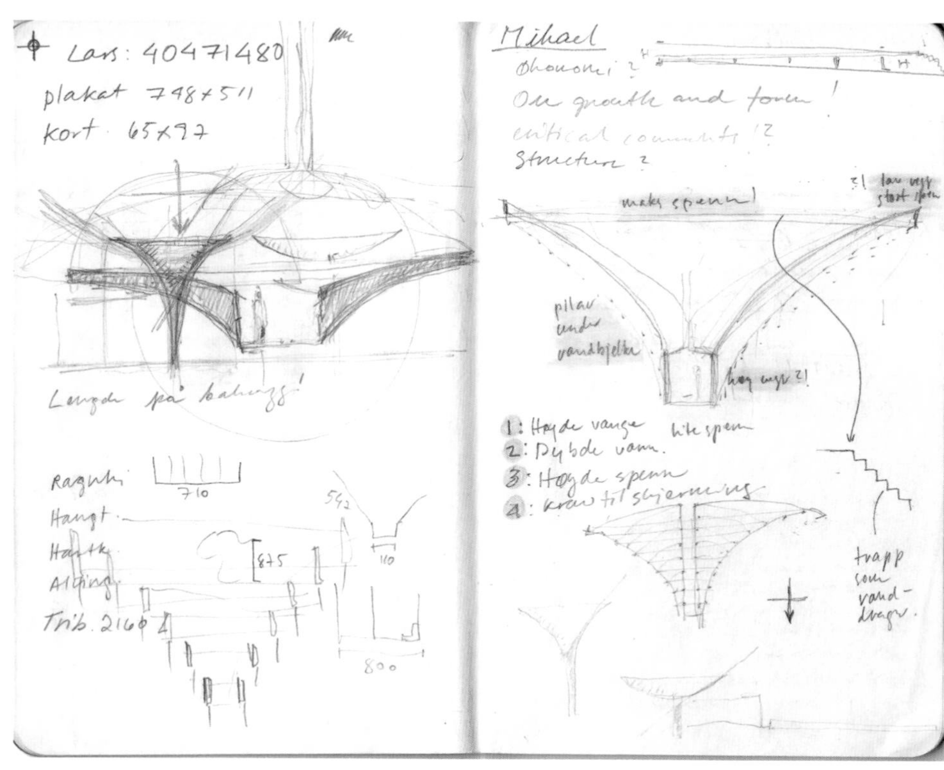
Lars: 40471480
plakat 748×511
Kort 65×97
Mihael
On growth and form!
Structure ?
maks spenn!
pilar
under
vandbjelke
1: Høyde vange
2: Dybde vann.
3: Høyde spenn
trapp
som
vand-
drager.
710
875
800

This project is a public pier into one of Lake Vänern's many delta bays in Karlstad, the capital of Värmland, a Swedish region where the timber and iron industries have a long history.

The municipality is developing new housing blocks in a former industrial part of the city. The pier is a public art project and as such connected to the municipal urban development. It is a prolongation of the central axis through the densely developed new area. It points south-westwards towards the natural reserve on the opposing shore.

The form of the pier allows many people to spend time on that which is a pier's most attractive part – the very end. The floor of the pier is slightly sloped towards a large public stair leading down to the water. Tall benches line the sides of the pier creating a space that is sheltered from wind and from the city view. Occasionally the regulated water of the lake will fill the space of the pier reaching all the way up to the promenade.

The pier is a concrete structure: pillars, slabs, floor and 160 m of prefabricated bench elements. The area of the pier is 1,500 m².

Credits and Data

Project title: Tullholmen Pier
Client: Karlstad Municipality
Program: Public Space
Location: Tullholmsviken, Karlstad, Sweden
Design: 2017–2018
Built: 2019
Architect: Manthey Kula
Team: Per Tamsen and Beate Hølmebakk, responsible architects
With Lars Holen
Structural engineer: Sweco as
Area: 1,500 m²

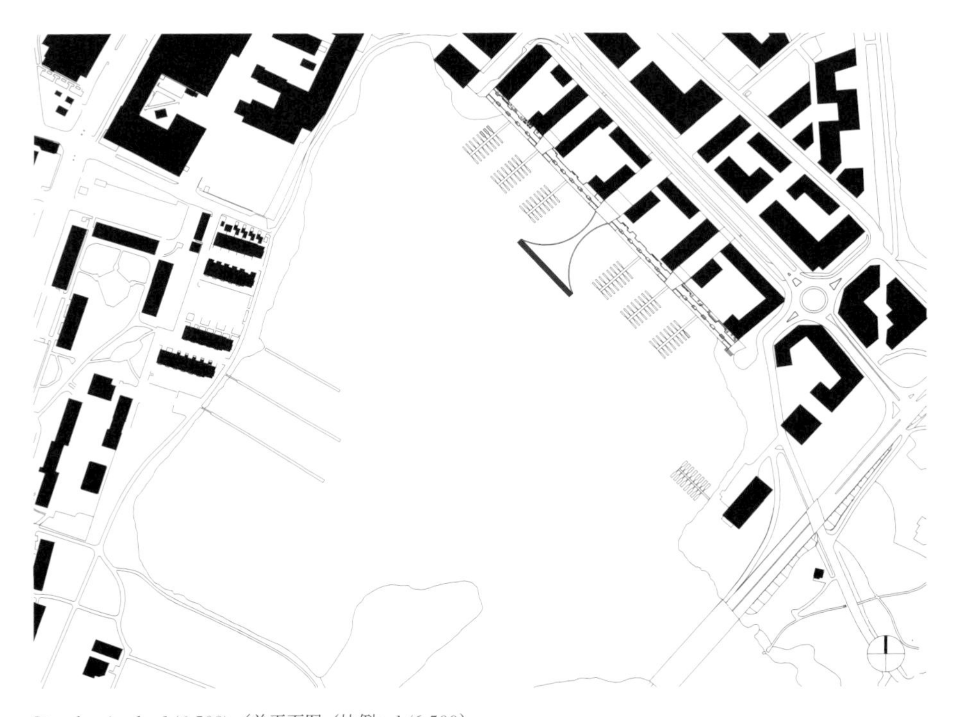

Site plan (scale: 1/6,500) ／总平面图（比例：1/6,500）

pp. 154–155: Photo of model showing tall benches that line the sides of the pier creating a space sheltered from the wind. Opposite, both: Esquisse sketches by the Architect. p. 159, above: CG images on the pier. Winter and summer. p.159, below: Section persepective drawing. Images on pp. 154–159 courtesy of Architect.

第154-155页：高背长椅模型。沿栈桥两侧而设的高背长椅创造出了一个挡风空间。对页，上图和下图：建筑师在研究阶段的手绘稿。第159页，上：栈桥冬天和夏天的效果图；下：剖面透视图。

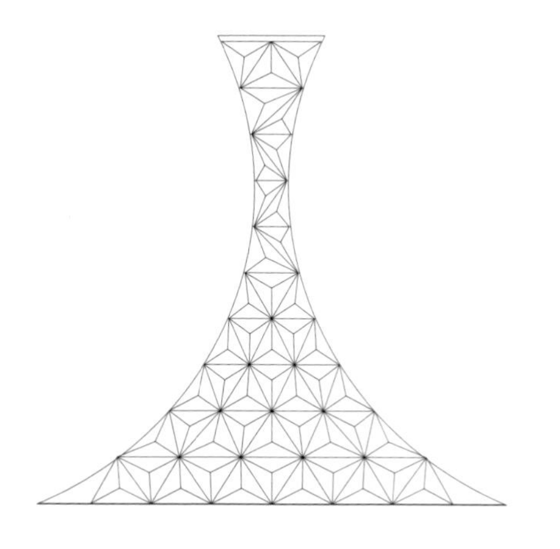

Drainage plan ／排水平面图

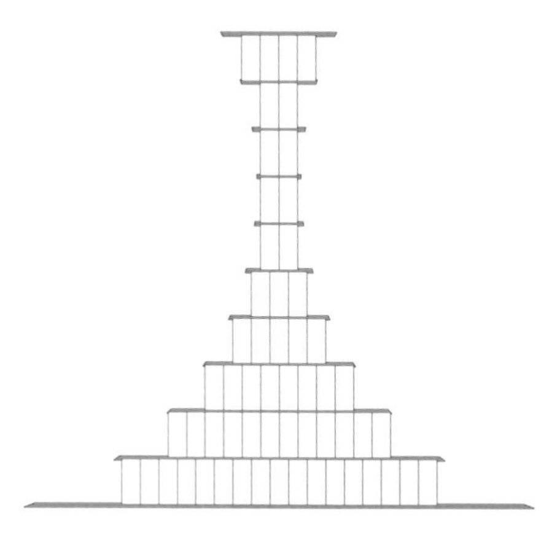

Beams and flooring elements plan ／梁和楼板平面图

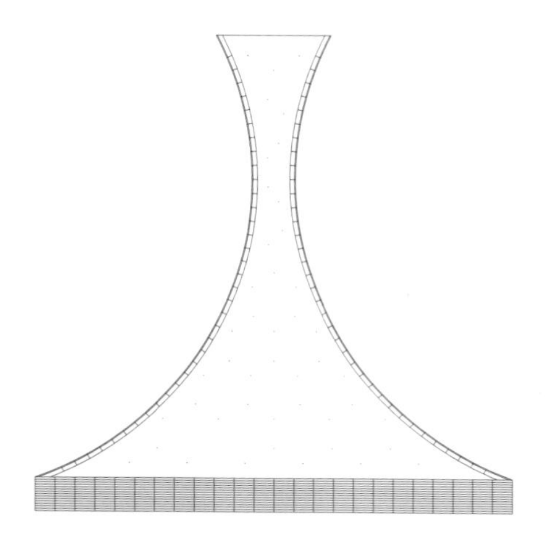

Benches and stairs plan ／长椅及台阶平面图

Insite edges plan ／内部边界平面图（比例：1/300）

本项目是一座公共栈桥，位于瑞典卡尔斯塔德市维纳恩湖的一个三角洲湾中。卡尔斯塔德是韦姆兰省的首府，韦姆兰是瑞典境内一个拥有悠久林业和钢铁工业历史的地区。

市政府正在开展一个坐落于原工业区的新建住宅规划。这座栈桥被作为一个与该规划关联的公共艺术项目，位于该高密度、新开发的区域中轴线延长线上，并指向西南方对岸的自然保护区。

栈桥的造型使得其尽头成了一个最具魅力的地方，吸引了众多游客。栈桥地面朝向一个巨大的公共台阶稍稍倾斜，台阶与下方水面相接。沿栈桥两侧而设的高背长椅创造出了一个隔绝风以及都市景色的空间。人工调节的湖水有时会填满栈桥两侧的空间，一直上涨到与步行道齐平的高度。

栈桥为通体混凝土结构，包括柱子、层板、地板和预制总长160米的长椅部分。栈桥的总面积1,500平方米。

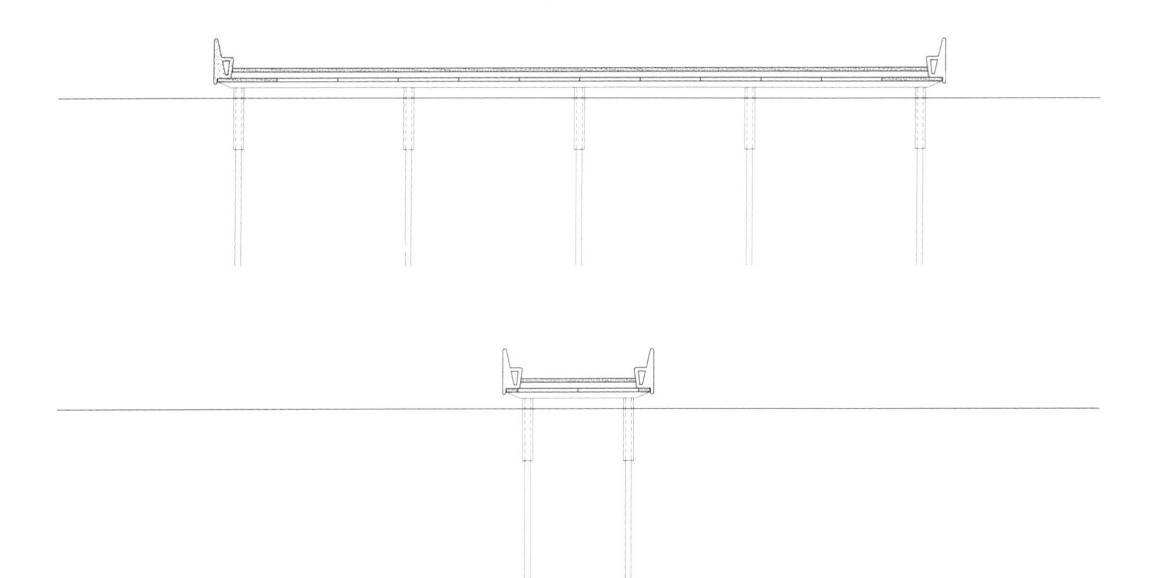

Sections (scale: 1/300) ／剖面图（比例：1/300）

Manthey Kula

Archipelago – Building from Solitude

2017

曼蒂·库拉

群岛——孤岛的建筑

2017

The projects' exact formal and spatial ideas are developed from speculation and impressions; as such this work explores the relationship between order, association and intuition.

The project was developed for the Frac Centre-Val de Loire on occasion of their Architectural Biennale Marcher Dans le Rêve d'un Autre. Constraints were used to create necessary resistance in an otherwise open process. They were:

- That the narrative source should be of exiles on islands

- That each project should be developed on one sheet of paper

- That the size of the scale models, hence buildings, should be limited to one unit of a measuring system used in the country of the protagonist at the actual time

- That all the projects should respond to specific grounds to be made in recycled Canadian oak

The work is generated from brief accounts of five individuals that experienced profound isolation spread out across history and across the map. The fates of these people are known. However their dreams – products of lived impressions, existential reasoning, and subdued thoughts – remain outside our reach. Of these dreams we can only speculate.

Credits and Data

Project title: Archipelago – Building from Solitude
Client: Frac Centre-Val de Loire
Location: Collection of the Frac Centre-Val de Loire, 88 Rue du Colombier, 45,000 Orléans, France
Design: 2016–2017
Architect: Manthey Kula
Team: Beate Hølmebakk and Per Tamsen, responsible architects with Loris Clara and Lars Holen

Opposite and this page: Model making process. Images on pp. 160–173 courtesy of the Architect.

对页和本页：模型制作的过程。

这个项目确切的形式和空间概念源于深入推演和特定印象，因此这里真正探讨的是游走在秩序、联想和直觉之间的体验关系。

这个项目是为法国卢瓦尔河谷中心的建筑双年展而准备的（双年展名为“漫步别人的梦中”）。项目本身是开放的，但也设置了以下限制作为必要的约束条件：

- 作为故事主人公必须是在岛上流亡的人。
- 每个项目都必须在一张纸上完成。
- 比例模型的大小，也就是建筑本身，必须使用主人公所处时代和所属国家的统一单位。
- 所有的项目都必须对应特定基地，并使用再生的加拿大橡木制成。

这个作品源自对5个经历了深刻孤独与隔绝的人的简短描述，他们属于不同历史时代，位于不同的地方。这些人的命运是众所周知的，然而有关他们生前的印迹、实际存在的证明和压抑的思想，仍然是我们可以探究的。对于这些梦境，我们只能推测。

阿森松岛
Acension Island
7° 56' 0" S, 14° 22' 0" W

圣尼古拉斯岛
San Nicolas Island
33° 15' 51" N, 119° 32' 20" W

索洛维茨岛
Solovets Island
65° 5' 0" N, 35° 53' 0" E

北兄弟岛
North Brother Island
40° 47' 53" N, 73° 53' 54" W

德朗盖岛
Drangey Island
65° 56' 11" N, 19° 40' 45" W

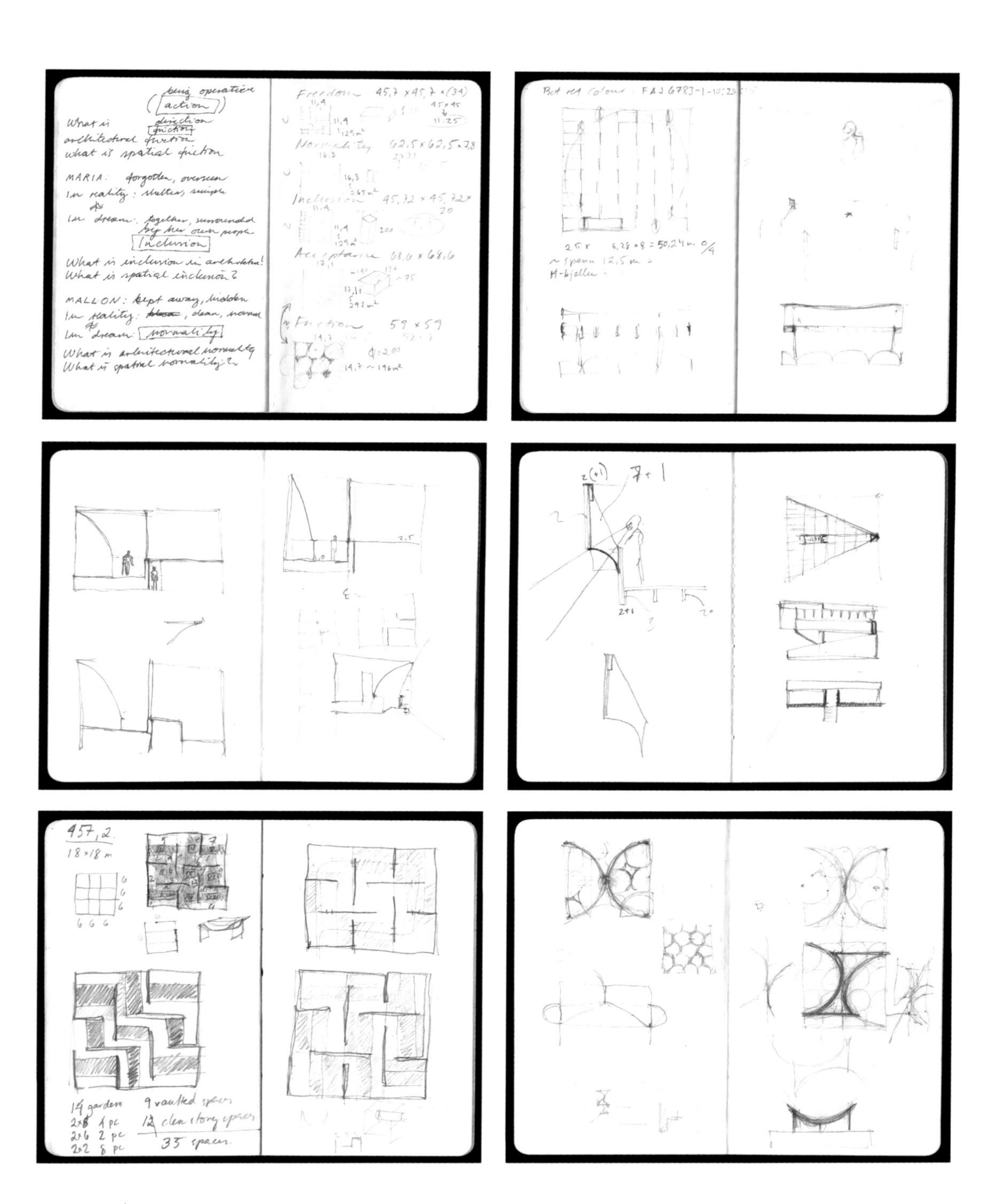

Opposite: Islands where people were exiled. This page: Sketches by the Architect.

对页：主人公们被流放的小岛。本页：建筑师的手绘。

Archipelago – Building from G

群岛——G的建筑

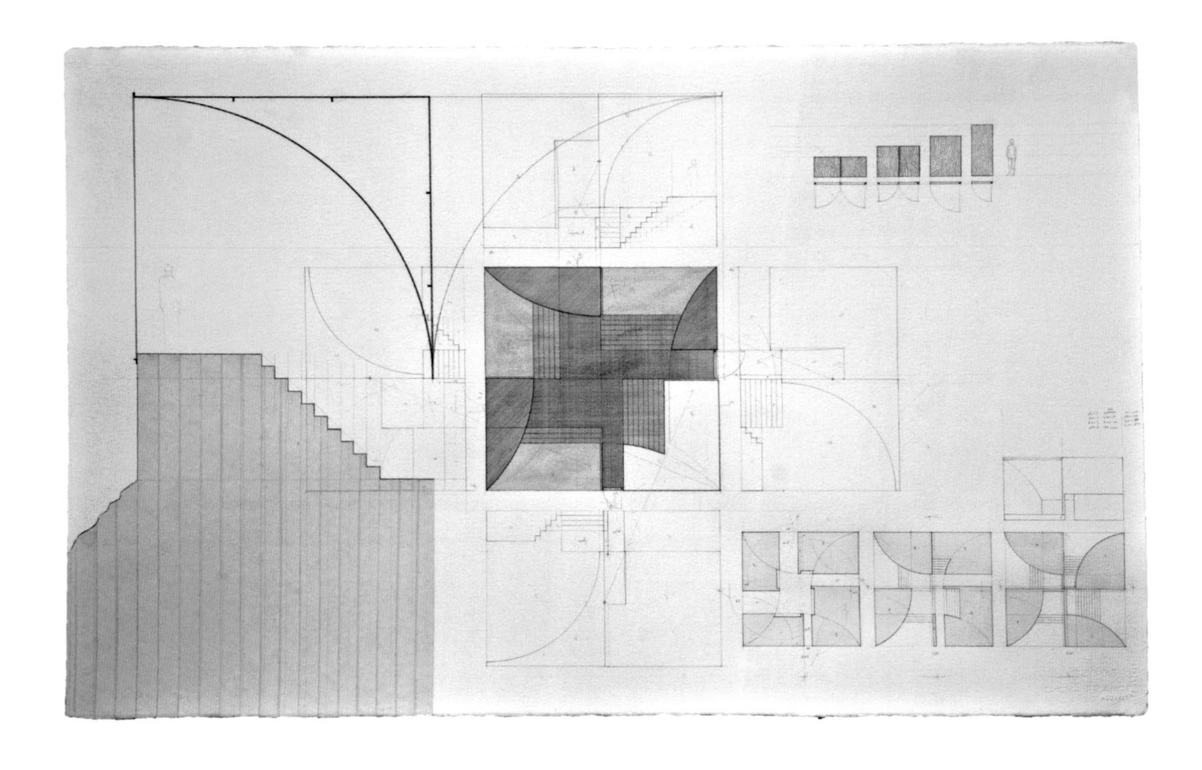

Grette Aasmundsson (isolated from 1045 to1050) of Iceland who withdrew to Drangey Island in the North Atlantic Ocean after being found guilty of arson:

Building from G: The cave and the cardinal directions.

There must have been dreams of Fear and Freedom, and a need to be in control.

冰岛人格雷特·艾斯蒙德森（1045年到1050年被流放）因纵火罪而被流放到北大西洋上的德朗盖岛上：

G的建筑：洞穴和基本的方向。

他一定有过关于恐惧和自由的梦，也有自控的需要。

Opposite: Sketches are developed on one sheet of paper. This page, left: Interior view of the model. Opposite, right: General view of the model.

对页：手绘全部在一张纸上完成。本页，左：模型的内部景观；右：模型全景。

Archipelago – Building from L

群岛——L的建筑

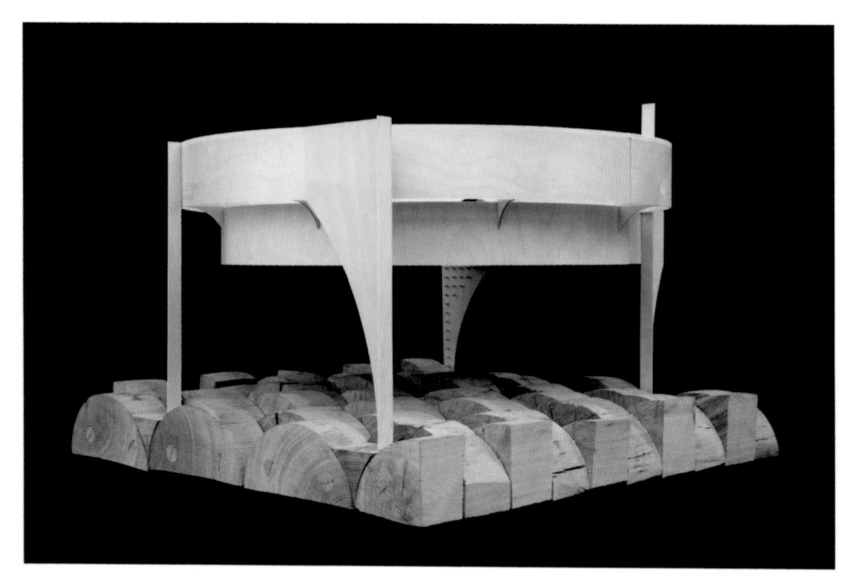

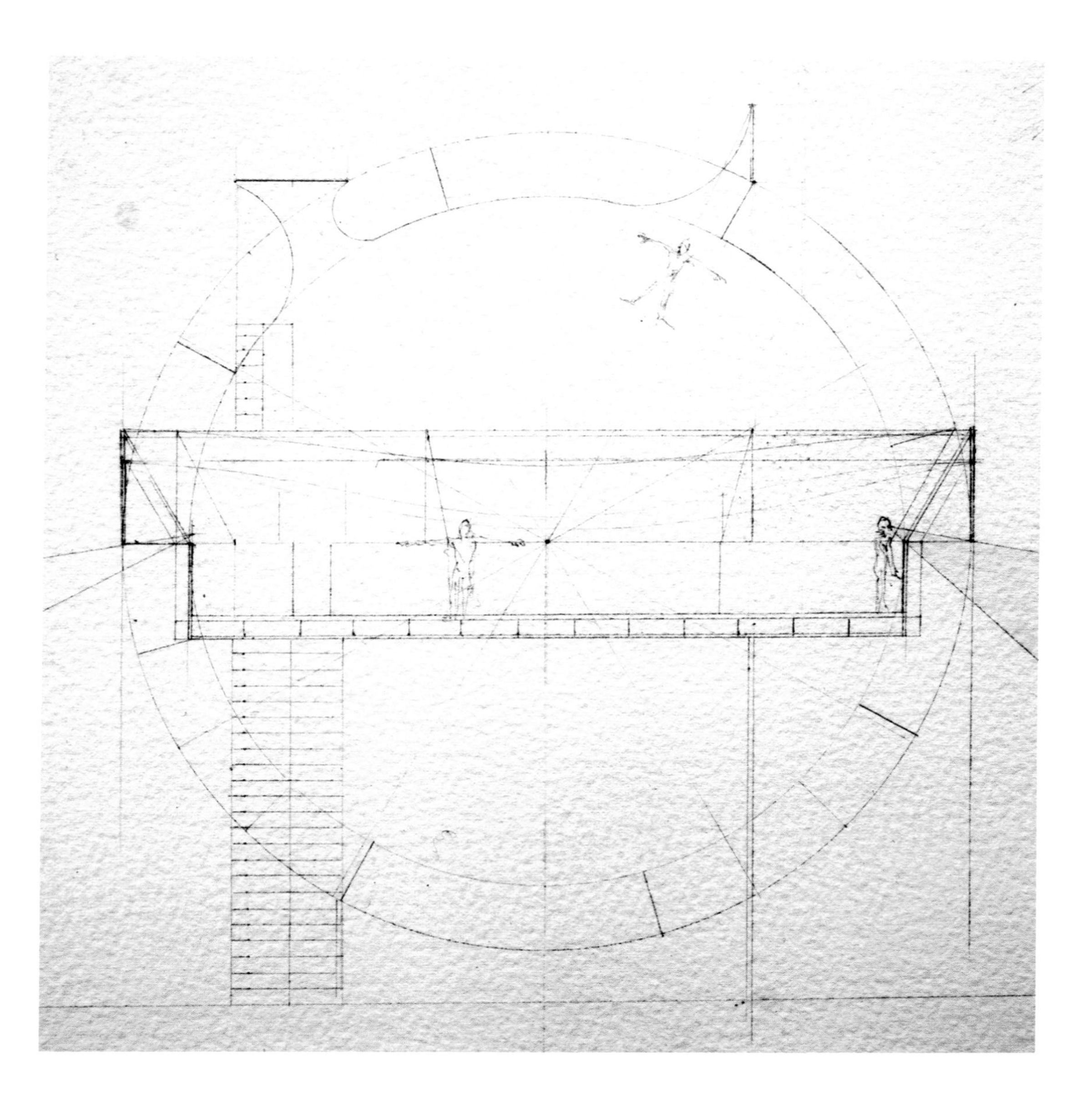

Leendert Hasenbosch (isolated in 1725) of Holland who was set ashore on Ascension Island in the South Atlantic Ocean after being revealed as a homosexual:

Building from L: The horizon and the vertical dimension.

There must have been dreams of Acceptance, and a yearning to surrender.

荷兰人利安德特·哈森博什（1725年被流放）因为其同性恋的身份被揭发，被流放到了南大西洋的阿森松岛上：

L的建筑：水平线和垂直面。

他一定有过关于被认同的梦想，以及对屈服的渴求。

Opposite, above: General view of the model. Opposite, below: View of the model from above. This page: Sketch with plan and section drawn in a single frame.

对页，上：模型全景；下：俯瞰模型。本页：同时表现平面和剖面的手绘。

Archipelago – Building from P

群岛——P的建筑

Petro Kalnyshevsky (isolated from 1690 to1803) of Ukraine who was exiled to the Solovets Island in the White Sea after having fallen from grace of the Russian Empire:

Building from P: The gravity and the cross.

There must have been dreams of Friction, and a feeling of gratefulness.

乌克兰人彼得罗·卡尔尼舍夫斯基（1690年到1803年被流放）因为遭到了俄罗斯帝国的迫害，被流放到位于白海的索洛维茨岛上：

P的建筑：重力和十字。

他一定有过关于分歧的梦，以及一种感恩的心情。

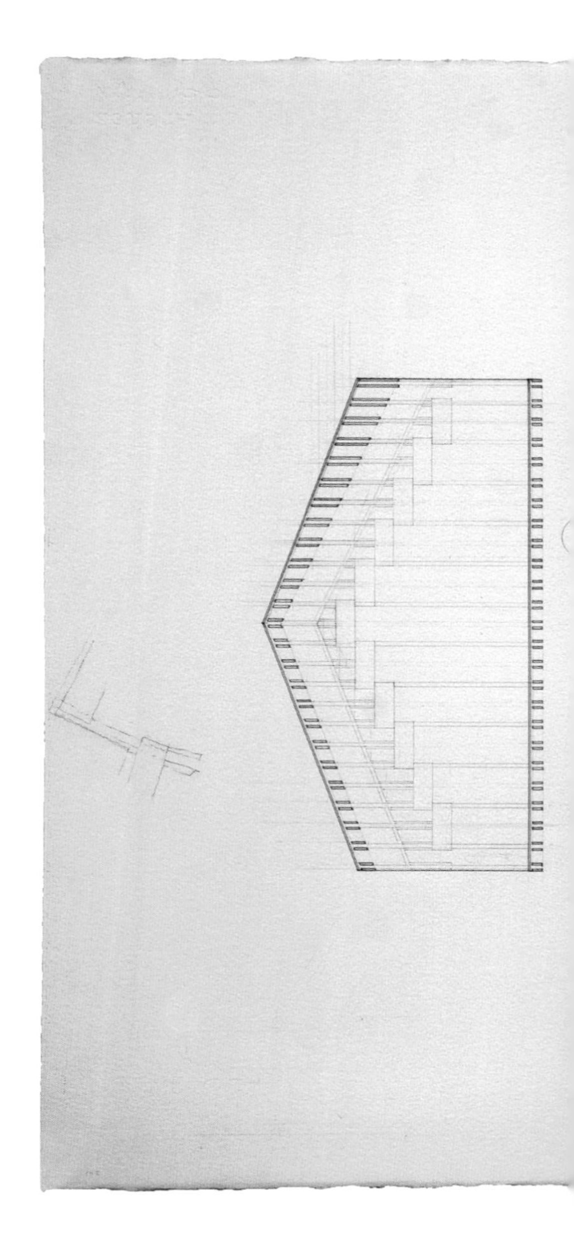

Opposite, above: General view of the model. Opposite, below: Top view of the model. pp. 168–169: A sheet of the sketch.

对页，上：模型全景；下：模型顶部。第168-169页：一张纸上的手绘。

Archipelago – Building from J

群岛——J的建筑

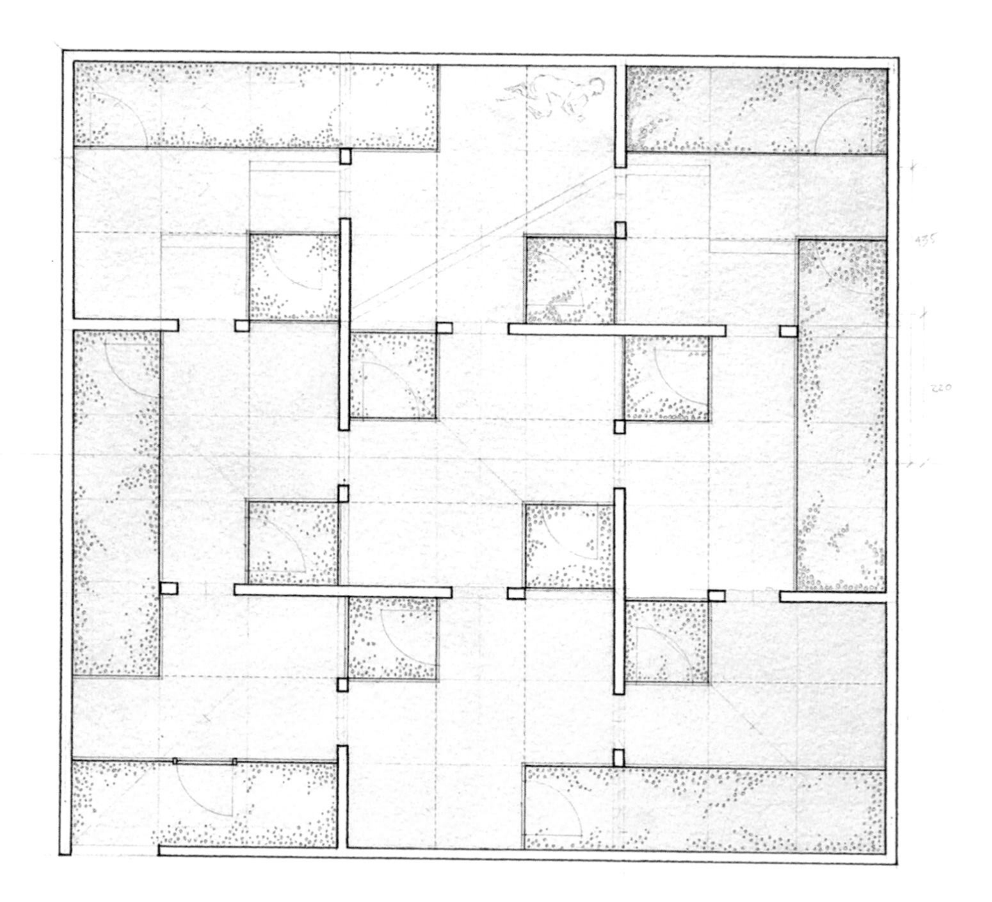

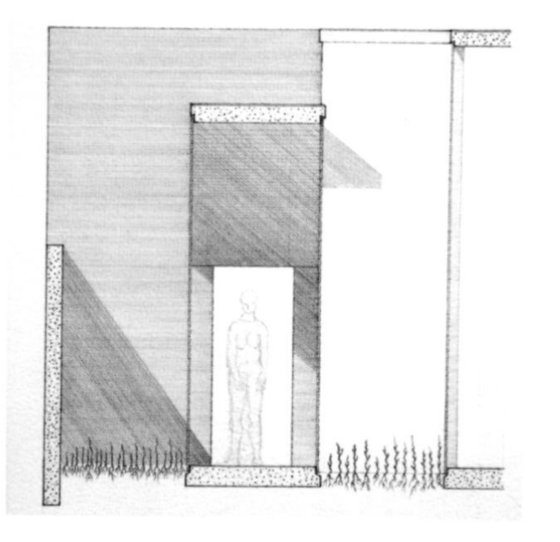

Juana Maria (isolated from 1811 to 1853) of the Nicoleños who was left on San Nicolas Island in the Pacific Ocean when Franciscan missionaries removed her tribe:

Building from J: Nine rooms and fourteen gardens.

There must have been dreams of Inclusion, and persistence.

尼科尔尼奥斯的胡安娜·玛丽亚（1811年到1853年被流放）因为自己的部落遭到方济各教会传教士们的排挤，被流放到了太平洋的圣尼古拉斯岛上。

J的建筑：9个房间和14个花园。

她一定有关于包容和坚持的梦。

Opposite, above: Plan drawing with nine rooms and fourteen gardens. Opposite, below left: Interior view of the model. Opposite, below right: Drwaing of the details. This page: View of the model from above.

对页，上：9个房间和14个花园的平面图；左下：模型内景；右下：细节部分图纸。本页：俯瞰模型。

Archipelago – Building from M

群岛——M的建筑

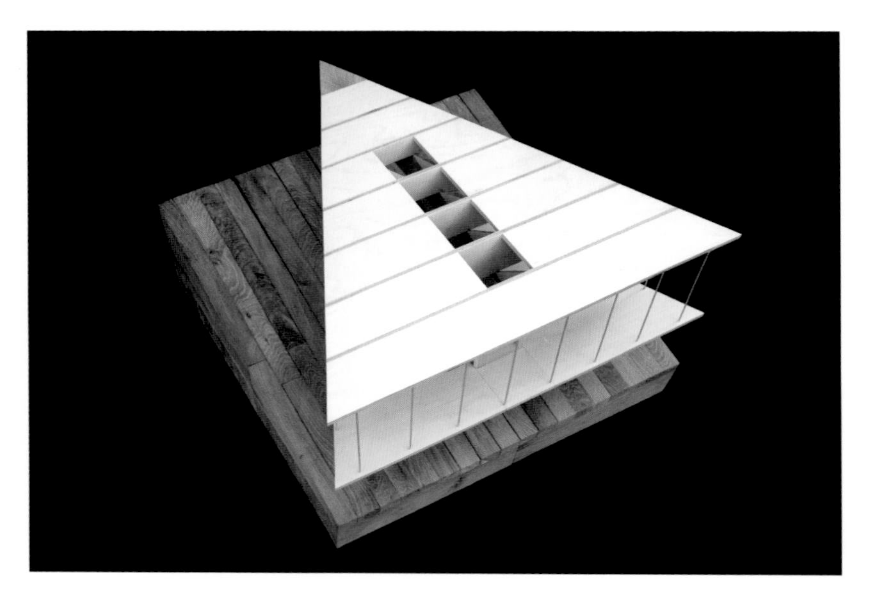

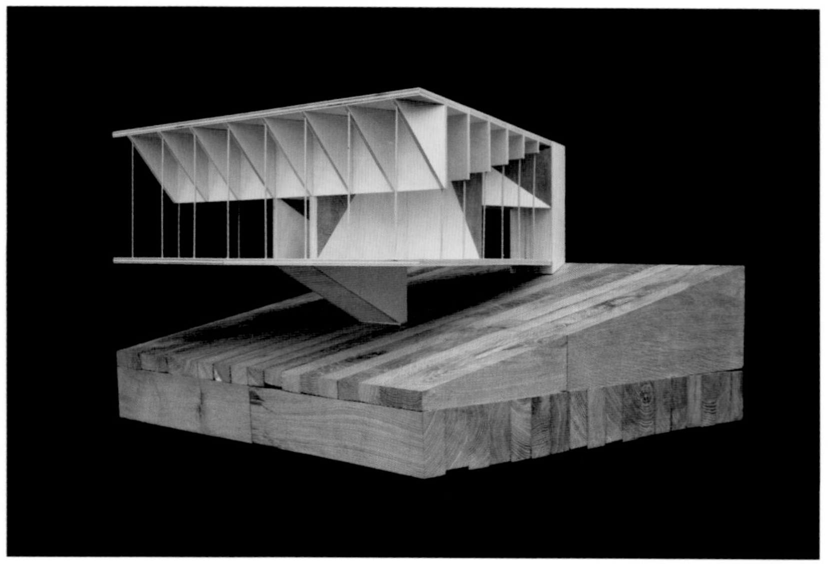

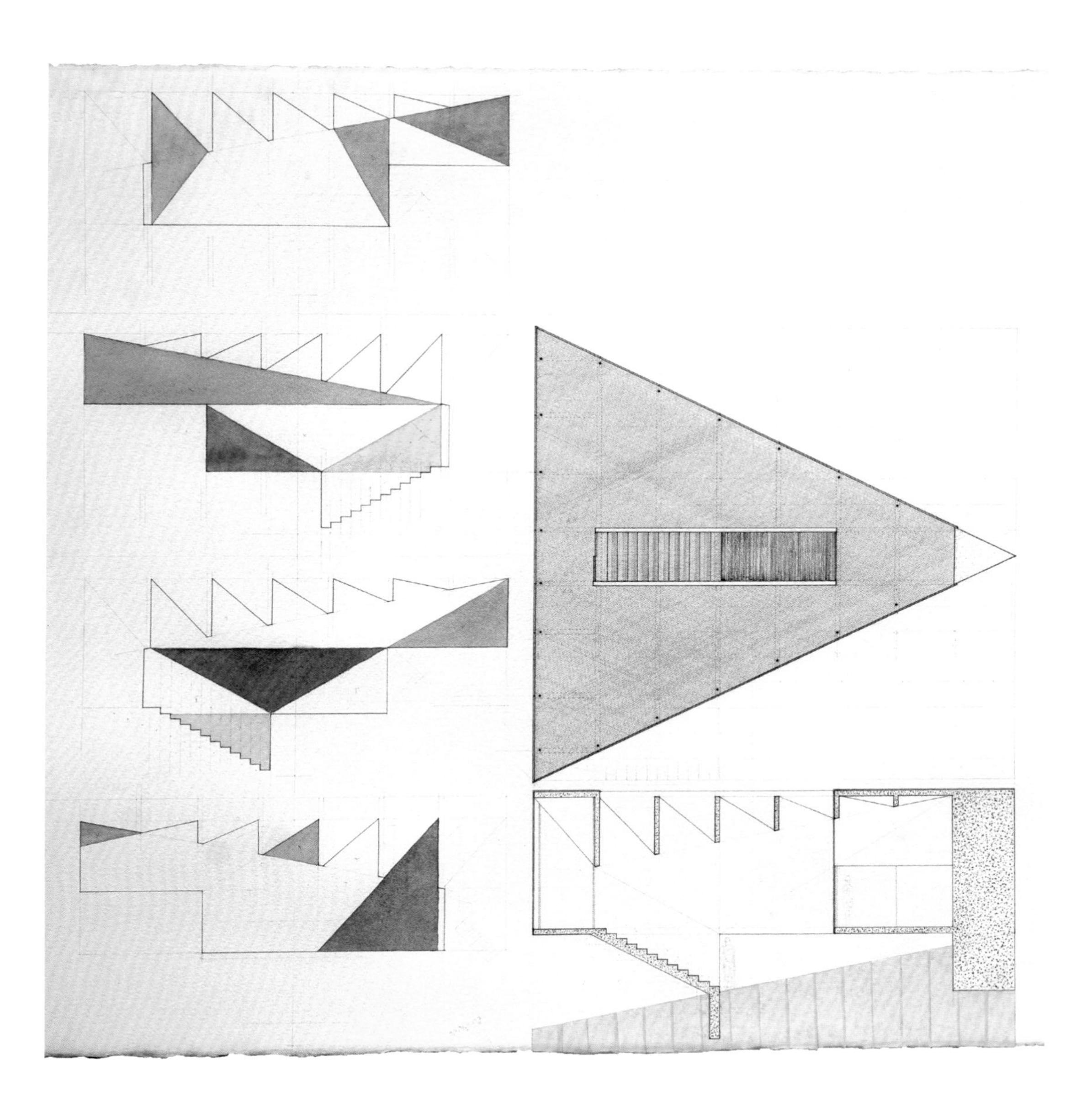

Mary Mallon (1869-1938) of Ireland who was sent to quarantine on North Brother Island in the East River, New York after having infected 22 people with typhoid fever through her work as a cook:

Building from M: The incline and the level.

There must have been dreams of Normality, and a concern about surface.

爱尔兰厨师玛丽·马龙(1869-1938) 因致22个人感染伤寒而获罪，后被流放到了纽约东河的北兄弟岛隔离检疫：

M的建筑：斜坡和水平面。

她一定有过正常生活的梦想，以及对表面的关注。

Opposite, above: View of the model from above. Opposite, below: General view of the model. This page: Drawing (part).

对页，上：俯瞰模型；下：模型全景。本页：（部分）图纸。

Atelier Oslo

Villa Holtet

Oslo, Norway 2015

奥斯陆工作室
霍尔特别墅
挪威，奥斯陆 2015

The starting point for this task was a typical challenge in Oslo: densification in an area of existing single family houses. The plot was a lovely old garden. It was important for us to preserve much of the garden for outdoor use, but also as a natural quality to the enjoyment of residents, both outside and inside. Therefore, the ground floor has a relatively limited footprint, while the larger upper floor cantilevers out, creating covered outdoor areas.

The house is broken up into smaller volumes to adapt to the relatively tight situation and the scale of the surrounding buildings. For the clients to finance the house, a part of the house is made as a separate unit to rent out. This area can easily be integrated into the house over time.

As an addition to the garden, and as compensation for the reduced view, the project creates an inner landscape, a sequence of rooms with varying scale and use, different degrees of transparency and privacy, changing views and light conditions. The central double-height living room is the heart of the house and connects all the rooms and areas. The room is surrounded in first floor by lobby, kitchen and dining room, and the garden with its various outdoor areas.

A staircase leads up to a gallery with access to the private rooms, bedrooms and bathrooms. The gallery is a casual place, and acts as an extension of the rooms. A large skylight provides varied light and shadow effects through the day.

The house structure is prefabricated wooden columns and beams. All structural parts are exposed in the interior of the project. The cantilevered rooms on the second floor hang from high beams under the roof. The beams have different dimensions, depending on the cantilever length. The columns have unique dimensions, adapted to their individual loads. It has not been the goal to standardize or clean up the construction, but rather let it be an organic result of the housing geometry.

A system that cannot immediately be read as a clear structure, but rather gives the qualities of a forest, where you can let your eyes wander along continuing branches. The project is carried out within a limited budget. An open and trusting cooperation between the builder, architect and contractor has nevertheless enabled a distinctive and elaborate house.

Credits and Data
Project title: Villa Holtet
Client: Private
Program: Single-family house with rent out apartment.
Location: Oslo, Norway
Built: 2015
Architect: Atelier Oslo
Contractor: Kjetil Eriksen
Area: 214 m²
Site Area: 600 m²

p. 175: Central double height living room is the heart of the house connecting to all rooms and spaces. This page: Cantilevered rooms on the second floor hanging from high beams under the roof. Opposite: Large skylight provides varied light and shadow effects through the day. Photo by Gunnar Sørås. pp. 178: Aerial view of the house. Photo by Gunnar Sørås. pp. 179: View of the house from the entrance driveway. p. 180: Split model. All photos on pp. 174–181 by Lars Petter Pettersen unless otherwise noted.

第 175 页：位于中央的两层通高起居室是住宅的核心，连接着所有的房间和区域。本页：固定在屋顶下的高梁上的二层的架空房间。对页：一扇巨大的天窗可以在一天中带来多变的光影效果。第 178 页：别墅鸟瞰图。第 179 页：从入口车道上看别墅。第 180 页：分割模型。

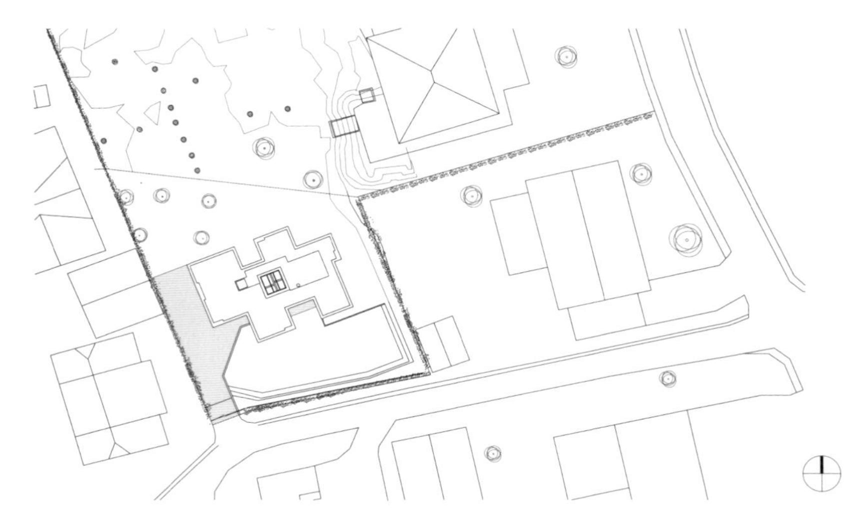

Site plan (scale: 1/800) ／总平面图（比例：1/800）

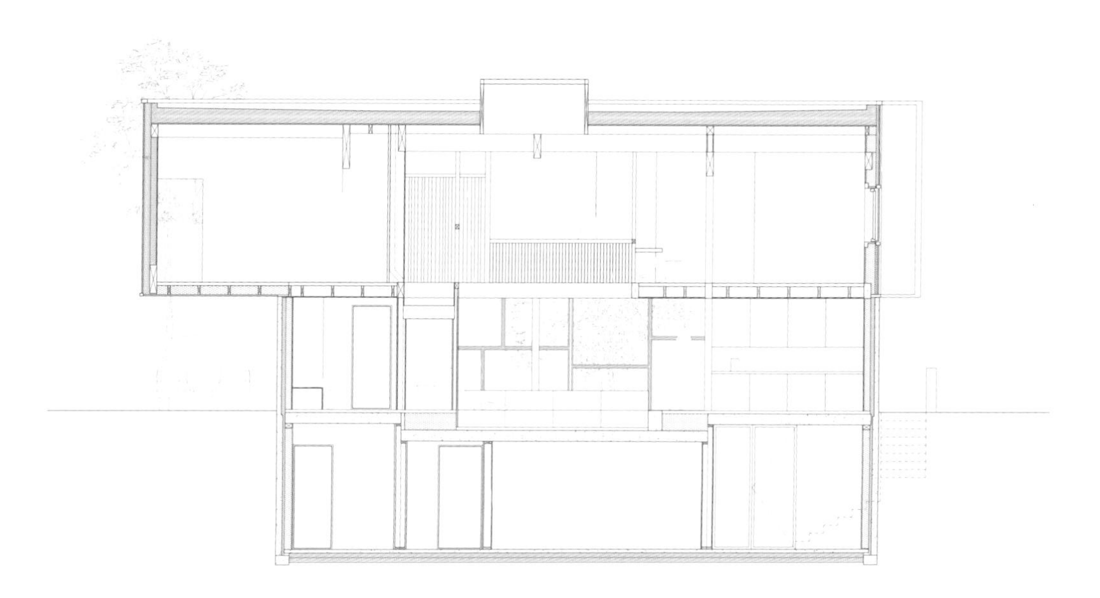

Long section (scale: 1/150) ／纵剖面图（比例：1/150）

这个项目的出发点是：增大奥斯陆现有单户住宅区的密度。该项目用地是一个古老而有魅力的花园。考虑到室外空间的利用，以及住户在室内外都能感受到高品质的自然生活，我们认为最大限度地保留现有的花园非常重要。因此，地上一层的面积变得相对有限，而面积较大的上层部分向外悬挑，形成了有覆盖的室外区域。

为了适应相对紧凑的布局和周围建筑的尺度，住宅被分解成多个更小的体量。为了让业主为住宅融资，住宅的一部分作为独立单元出租，而这个部分以后也很容易融入住宅主体之中。

作为对花园的补充，以及对景观减少的补偿，这个项目修建了一处室内景观，并设置了一系列不同尺度和用途的房间，营造了不同程度的透明性和私密性，以及多种视角和光线场景。位于中央的两层通高起居室是住宅的核心，连接着所有的房间和区域。起居室的一层部分被门厅、厨房、餐厅以及带有各种室外空间的花园所包围。

楼梯通向一条走廊，而经过走廊可以到达私人房间、卧室和卫生间。走廊可以看成房间的延伸，是一个休闲放松的空间。一扇巨大的天窗可以在一天中为住宅带来多变的光影效果。

房屋结构由预制的木柱和木梁构成。所有结构部件都暴露在住宅的内部空间里。二层的架空房间固定在屋顶下的高梁上。梁的截面随着悬挑的长度而变化。每根柱子根据承重的不同拥有各自独特的截面。项目的目标并不是要对结构进行标准化处理或是整顿，而是让它成为建筑几何的有机成果。

这个系统不能立即被解读为一个明快的结构，但空间会被赋予森林般的质感，你可以让你的目光流连在连续的树枝之间。项目的预算有限。尽管如此，建筑公司、建筑师和业主之间坦诚而互信的合作使这座独特而精致的住宅得以落成。

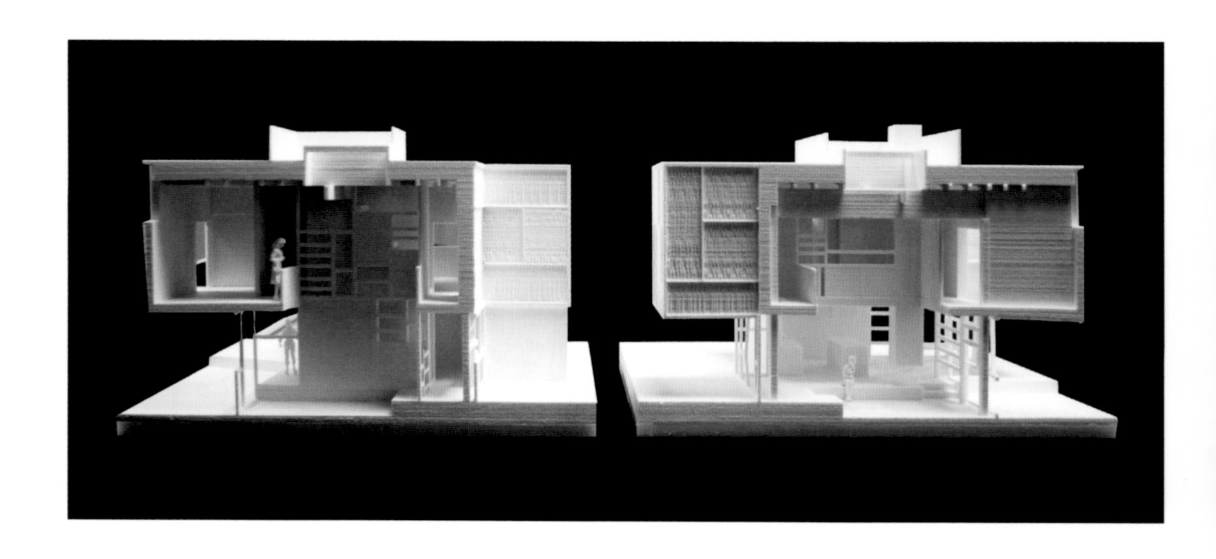

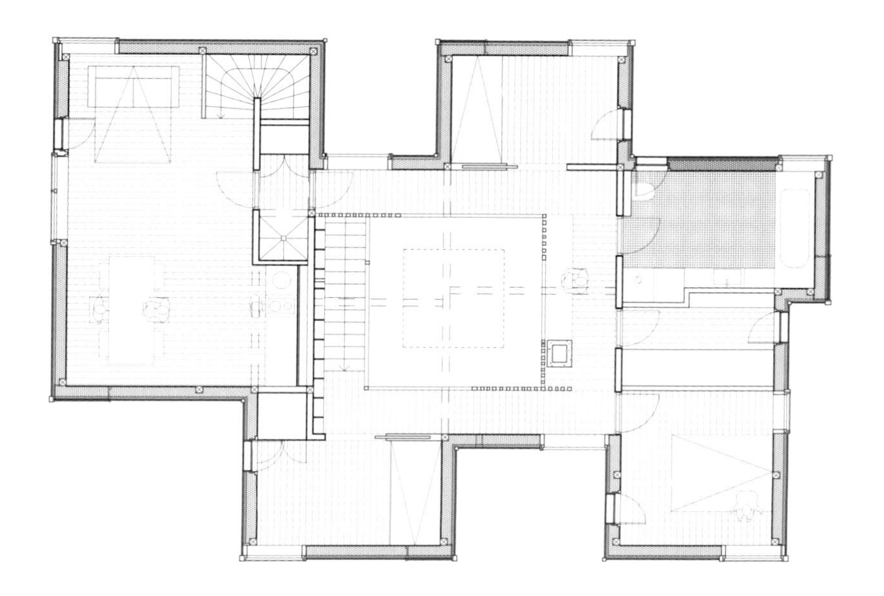

Second floor plan ／二层平面图

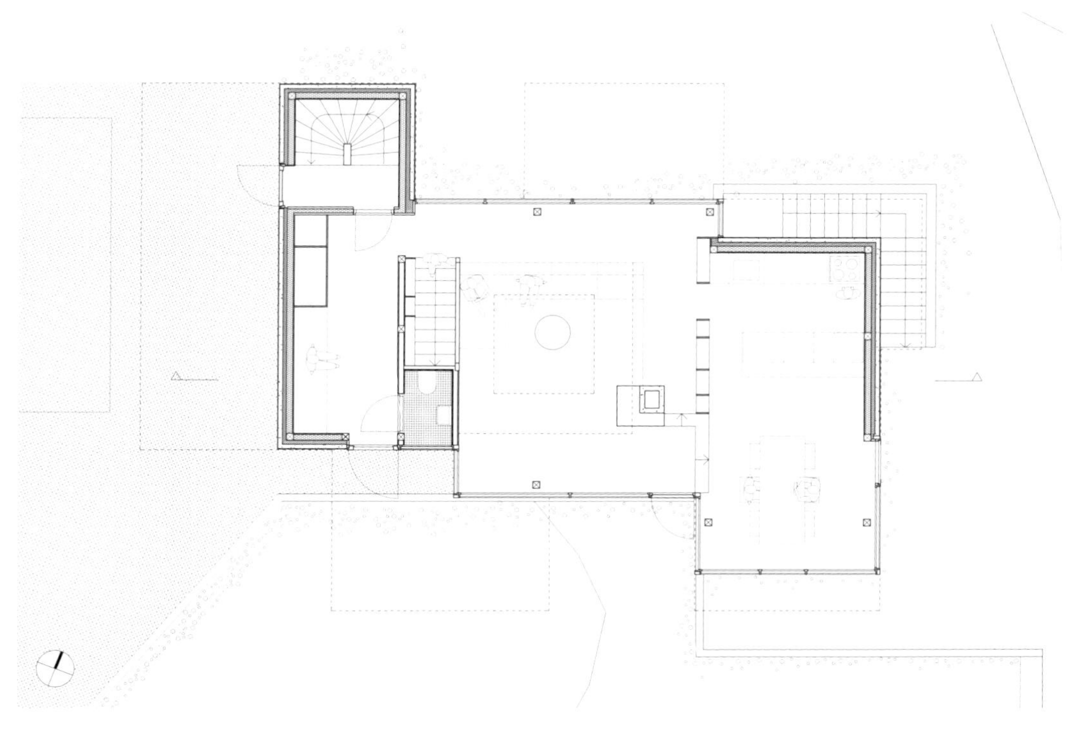

Ground floor plan (scale: 1/200) ／一层平面图（比例：1/200）

Atelier Oslo

House on an Island

Telemark, Norway 2018

奥斯陆工作室
岛上住宅
挪威，泰勒马克 2018

The small house is situated on an island on the south coast of Norway. The site is characterized by smooth and curved rocks that go down towards the ocean. The house is built for two artists that wanted a house for contemplation and working.

You enter the project from the backside walking along a small hill. The entrance is a stair that goes through the building taking you to the entrance on the front side. The entrance sequence marks a transition and prepare you for life on the island.

The topography of the site was carefully measured to integrate the rocks into the project. Concrete floors on different levels connect to the main levels of the topography and create a variety of different outdoor spaces. The concrete floors and stairs dissolve the division of inside and outside. The interior becomes part of the landscape and walking in and around the cabin gives a unique experience, where the different qualities from the site become part of the architecture. From the concrete floors, kitchen, bathroom and fireplace grow up to serve the inhabitants like furniture that sits on the rock.

A prefabricated timber structure is placed on the concrete floors and finally a light wood structure covers the cabin to filter the light and direct the views. The wood structure has a depth that creates a play of shadows through the day and a calm atmosphere resembling the feeling of sitting under a tree.

A small annex creates a fence towards the neighbor building and another sheltered outdoor space.

All exterior wood is Kebony, which is a special heat- treated wood that will turn grey and require no maintenance.

Credits and Data

Project title: House on an Island
Client: Private
Program: Summer house
Location: Skåtøy Island, Telemark, Norway
Built: 2018
Architect: Atelier Oslo
Consultant: Bohlinger + Grohman Ingenieure
Contractor: Admar
Area: 70 m²
Site Area: 700 m²

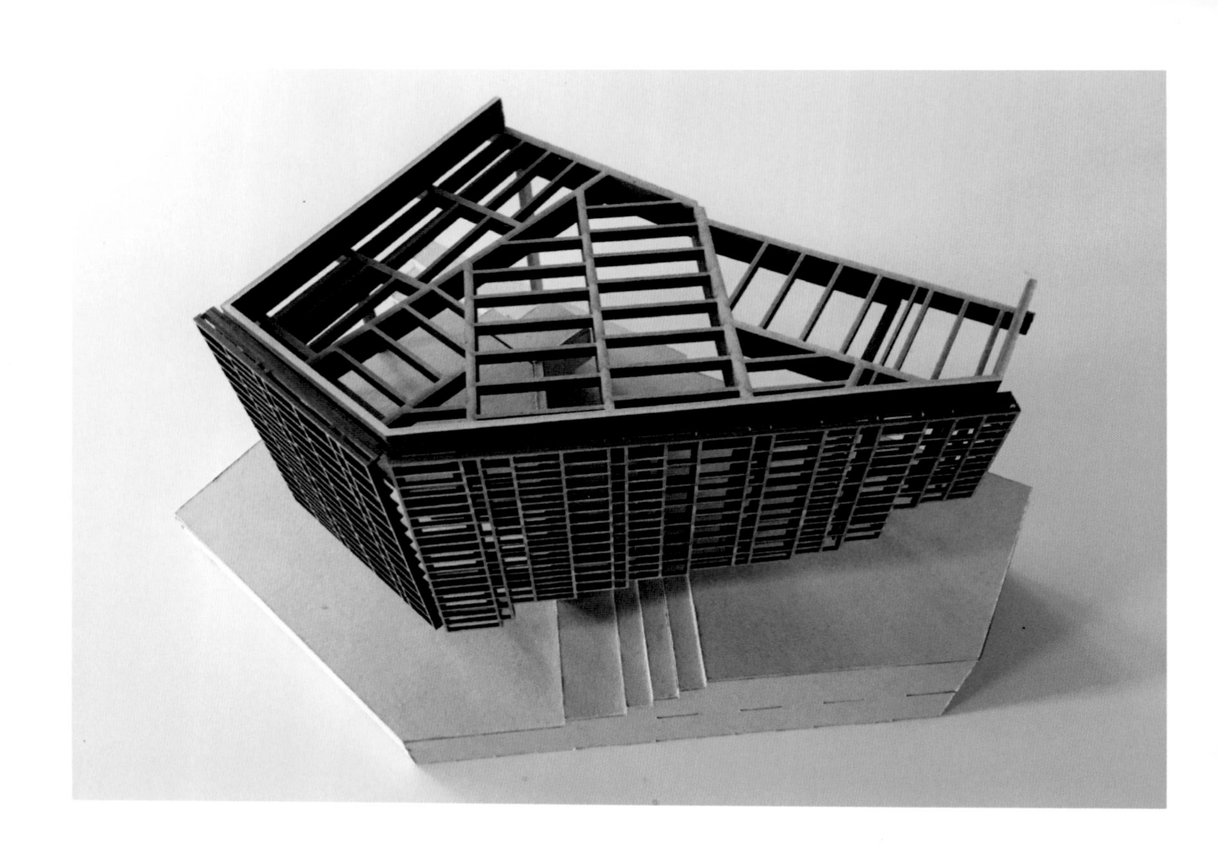

p. 183: External light wood structure filters light entering the house. All photos on pp. 182–191 by Nils Vik unless otherwise noted. This page: Model of the house frame. Opposite: The wood structure has a depth that creates a play of shadows through the day. Photo by Charlotte Thiis-Evensen.

第 183 页：被外部轻型木结构过滤后进入房屋内的光线。本页：房屋结构模型。对页：纵深的木结构可以在白天营造出光影交替的效果。

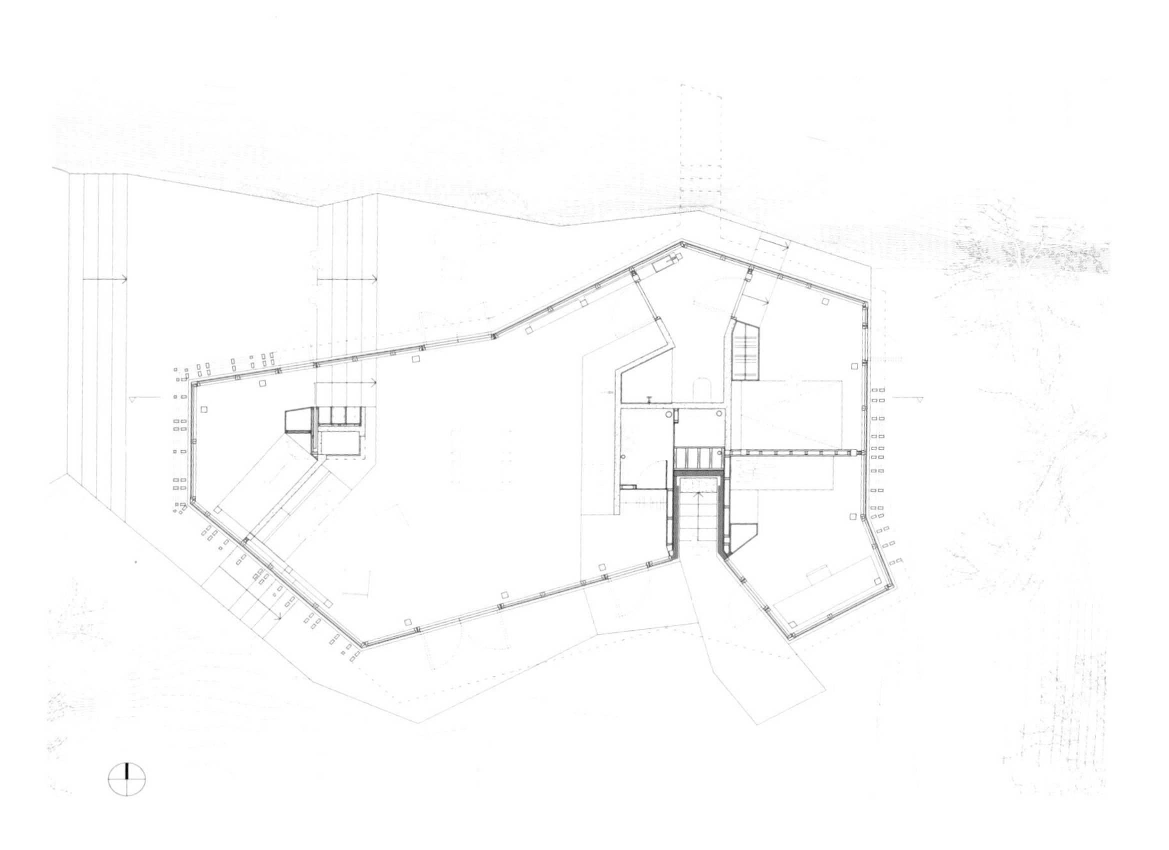

Site plan (scale: 1/150) ／总平面图（比例：1/150）

pp. 186–187: The use of timber on both its interior and exterior gives an atmosphere of sitting under a tree. Opposite: Exterior overall view of the house against the coast.

第186-187页：室内外均使用木材，营造出一种坐在树下的氛围。对页：面朝大海纵览建筑的外观。

这座小型住宅坐落在挪威南海岸的一个岛上。光滑而圆润的岩石向下延伸进入大海，赋予这个地块独特的个性。这座住宅是为两位艺术家建造的，他们希望这座房子可用于自己沉思和工作。

访客可从后方沿一座小山步入住宅之中。一段贯穿整座建筑的楼梯将访客带至正面入口。入口序列也是一个过渡，好让访客为进入岛上生活空间做好心理准备。

为了使岩石与建筑融为一体，现场地形经过了仔细勘测。不同高度的混凝土楼板连接到地形的主要水平面，创造出多样的室外空间。混凝土楼板和楼梯消解了室内外的分界。室内空间成了室外景观的一部分，穿行于小屋内外，人们可以获得一种独特的体验，基地不同的特质成了建筑的一部分。设置在混凝土楼板上为住户提供服务的厨房、卫生间和壁炉仿佛都是直接从岩石中生长出来的家具一样。

一个预制的木结构先被安装在混凝土楼板上，然后覆盖在整座小屋外部，起到过滤光线和引导视线的作用。纵深的木结构可以在白天营造出光影交替的效果，以及一种仿佛置身大树下的平静气氛。

一个小的附属结构在邻近的建筑和另一个带有遮蔽结构的室外空间之间形成了一个好似栅栏的建筑元素。

所有外部木材都使用一种名叫科博尼的材料。这是一种特殊的热处理木材，随着使用将逐渐变成灰色，并无须任何维护。

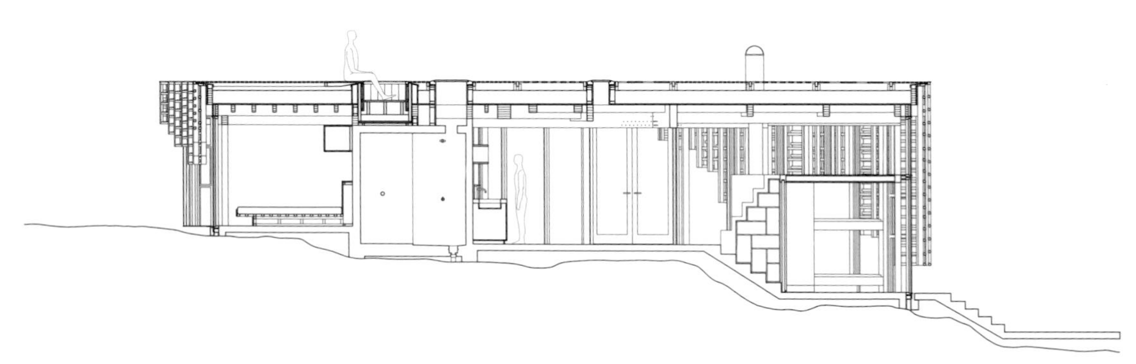

Long section (scale: 1/150) ／纵向剖面图（比例：1/150）

Opposite, above: Details of the external light wood structure. Opposite, below: View between the house and the site. Concrete floors on different levels connect to the existing levels of the site's choreography.

对页，上：外部轻型木结构的细节；下：建筑和场地之间的景色。不同高度的混凝土楼板与场地现有的不同高度的平面相连。

Atelier Oslo

Cabin Norderhov

Krokskogen, Norway 2014

奥斯陆工作室

诺德霍夫小屋

挪威，皇冠森林 2014

This page: View looking out into the landscape. Photo by Jonas Adolfsen. Opposite: Terraced steps leading into the bedroom. All photos on pp. 192–201 by Lars Petter Pettersen unless otherwise noted.

本页：屋外的景观。对页：通往卧室的台阶。

Credits and Data
Projec title: Cabin Norderhov
Program: Cabin
Location: Krokskogen, Norway
Built: 2014
Architect: Atelier Oslo
Contractor: Byggmester Bård Bredesen
Area: 70 m²
Site Area: 500 m²

pp. 194–195: View of the fireplace and dining room with views opening out into the landscape. This page, left: Curved walls and ceilings form continuous surfaces clad with 4mm birch plywood. This page, right: Gap through the walls brings natural light into the interior. Opposite: Floor details around the fireplace.

第194-195页：壁炉和餐厅，以及望向室外景观的景象。本页，左：曲面的墙壁和天花板形成连续的表面，其上覆盖着4毫米厚的桦木胶合板；右：光线沿墙壁上的缝隙进入室内。对页：壁炉周围地板的细节。

The project is located in Krokskogen forests, outside the town of Hønefoss. Its location on a steep slope gives a fantastic view over the lake Steinsfjorden.

The site is often exposed to strong winds, so the cabin is organized around several outdoors spaces that provide shelter from the wind and receives the sun at different times of the day.

The interior is shaped as a continuous space. The curved walls and ceilings form continuous surfaces clad with 4mm birch plywood. The floor follows the terrain and divides the plan into several levels that also defines the different functional zones of the cabin. The transitions between these levels create steps that provide varies places for sitting and lying down.

The fireplace is located at the center of the cabin, set down in the floor of the main access level. This provides the feeling of a campfire in the landscape. Seen from all levels in the cabin, you can enjoy the fireplace from far away or lying down next to it.

Large glass walls are located in the living and dining areas. The frames of the glass are detailed carefully to avoid seeing it from the inside. This creates a more direct relationship with the nature outside.

Outside, the cottage has a more rectangular geometry and the walls and roofs are covered with 20mm basalt stone slabs laid in a pattern similar the ones often used for traditional wooden claddings in Norway.

The lodge consists mainly of prefabricated elements. The main structure is laminated timber completed with a substructure of Kerto construction plywood. The Kerto boards are CNC milled and defines the geometry both externally and internally. The cabin is supported by steel rods drilled directly into the rock, supplemented with a small concrete foundation under the fireplace for stabilization.

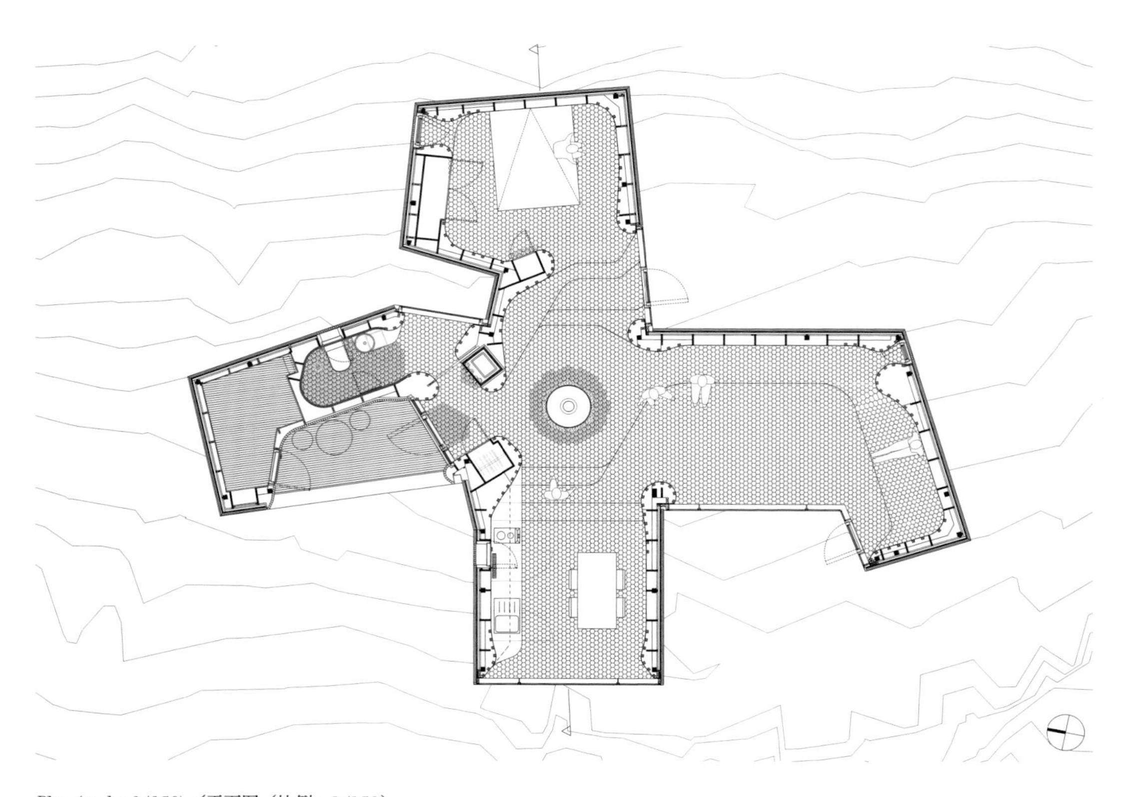

Plan (scale: 1/150) ／平面图（比例：1/150）

This page: Walls and roofs are covered with 20mm basalt stone slabs, similar to the traditional wooden claddings in Norway. Photo courtesy of the Architect.

本页：墙体和屋顶都覆盖着20毫米厚的玄武岩石板，石板的表面处理方式与挪威传统的木制表层非常类似。

这座小屋坐落在赫讷福斯镇郊外皇冠森林里的一个陡峭斜坡上，从小屋可以俯瞰屋外斯坦斯峡湾湖的湖光山色。

这片场地因经常暴露在强风中，所以在小屋周围设有若干个室外空间，强风来袭时，它们就是这里的避风所。此外这些空间能在一天中不同时间接受阳光的照射。

小屋内部被塑造成一个连续的空间。弯曲的墙壁和天花板形成连续的表面，其上覆盖着4毫米厚的桦木胶合板。地板跟随地形，被设计为若干水平面，水平面也定义了小屋的不同功能分区。这些水平面之间的交接处产生的台阶，为住户提供了多样的坐卧休息场所。

壁炉位于小屋的中心，设置在主要通道层的地板上。这样的设计给人一种在自然环境中营火的感觉。在小屋的各个水平面都能看到壁炉，住户既可以从远处欣赏它，也可以直接躺在它旁边。

客厅和餐厅装有大面积的玻璃幕墙。玻璃框架经过仔细设计，尽量避免在室内被看到，从而带来一种与室外大自然更强烈的联系感。

从室外看，小屋的几何造型更偏向矩形，墙体和屋顶都覆盖着20毫米厚的玄武岩石板，石板表面的处理方式与挪威传统的木制表层非常类似。

小屋通体几乎都由预制构件组成。主体结构由薄板的层压木材制成，底部结构则由一种名叫克尔托（CNC）的建筑胶合板构成。克尔托板由数控机床进行加工并定义了小屋内外的几何造型。整座小屋由直接钻入岩石的钢筋支撑，并在壁炉下补充了一个小混凝土地基以增加其稳定性。

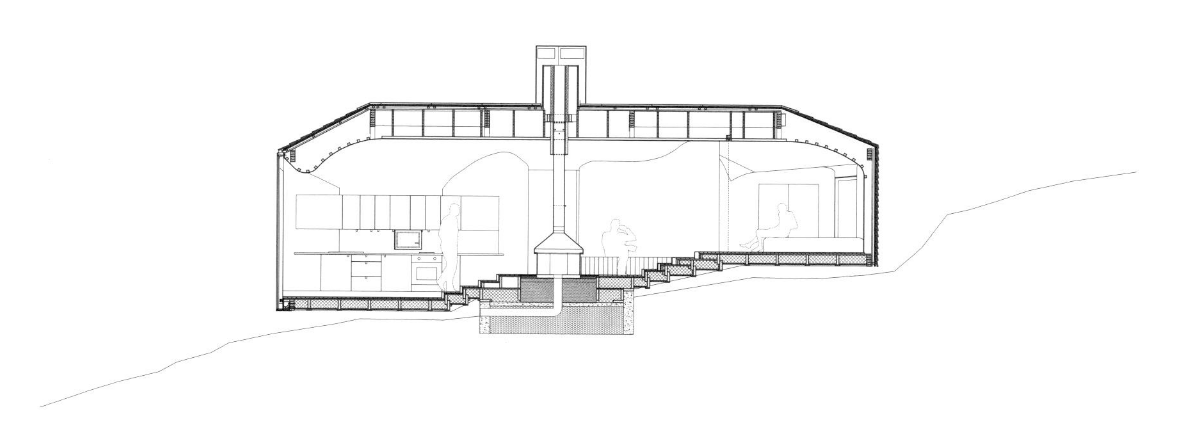

Section (scale: 1/150)／剖面图（比例：1/150）

Opposite: Exterior view from Lake Steinsfjorde.

对页：从斯坦斯峡湾湖看向建筑。

Atelier Oslo

Sentralen

Oslo, Norway 2016

奥斯陆工作室

森特拉伦

挪威，奥斯陆 2016

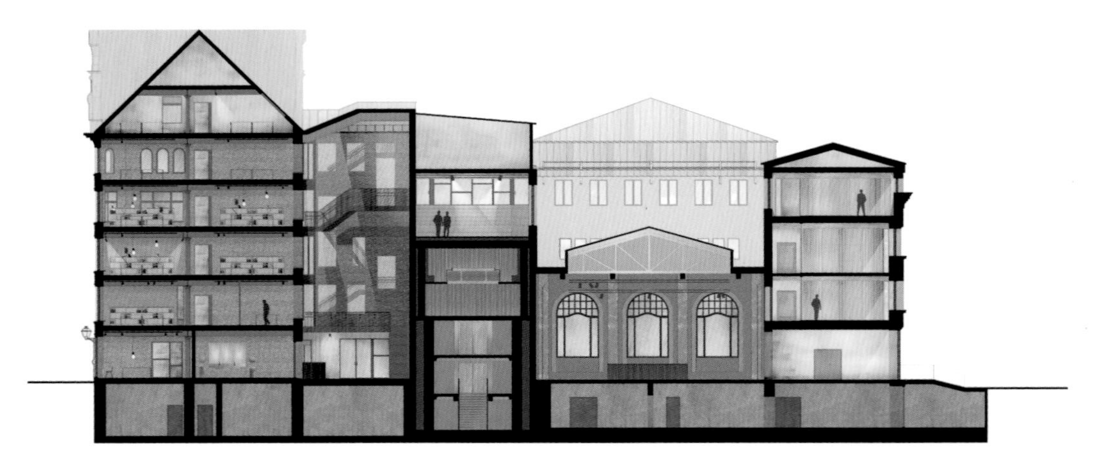

Section (scale: 1/500) ／剖面图（比例：1/500）

Credits and Data

Project title: Sentralen
Client: Sparebank Stiftelsen - Sparebank Foundation
Program: Cultural center. House for social innovation and cultural production
Location: Oslo, Norway
Built: 2016
Architect: Atelier Oslo and KIMA
Consultants: Fokus Rådgiving
Contractor: AF
Area: 13,000 m²
Site area: 2,700 m²

The new Cultural Center Sentralen occupies Christiania Sparebank's old bank building (1899) and the adjacent office building (1900) which was acquired by the Sparebank Foundation in 2007. In 2012 Atelier Oslo and KIMA won an open architectural competition, where the proposal suggested a work process and the description of it, without any drawings. The work process envisioned the design of a new cultural center, and the implementation of such a reconstruction. The plan layout and content of the project at this point were completely open.

In the process of transforming the old buildings, the emphasize were put on the buildings' original qualities and reuse of them in the best possible way. It was exciting work to find out which qualities could be preserved, and which lay hidden behind many layers of carpeting, plaster and ceilings. Many container loads of debris were carted out through the process. Several structural changes in the protected facades and interiors were made to open the former bank buildings and create an inviting and welcoming building. As the old buildings were never designed to be connected, one of the biggest new changes was to link the two buildings into a new and overall unity. A new entrance was created to Øvre Slottsgate and an elevator and staircase were added linking all parts of the building together. A new roof over the former backyard creates an new indoor square, "the winter garden", which has become a natural focal point for visitors and residents.

In the process, it was important to add the fewest possible new elements; only fire and sound partitions, universal design elements, acoustic absorbers and technical infrastructure were added. Many of the old surfaces have been preserved, while certain special rooms have been restored and returned to their original configuration. This contrast between the rustic and the refined gives the building a distinctive atmosphere.

A large working model of the buildings was built to be used for everyone involved to take part in the process of the project. Different solutions for new circulation, decor and design were tested and discussed. In this model, one could envision new uses of the various rooms, and what relation these rooms should have with each other.

The project laid particularly strong emphasis on flexible and functional solutions. All rooms are multifunctional and can be used for various activities. Meanwhile, the number of work stations can easily be adjusted. All technical installations are exposed in the ceiling and can be easily modified. The project is built to withstand changes over time.

The buildings have a high preservation value. A good dialogue with the Byantikvaren (Municipal Cultural Heritage Management Office) in Oslo was established early in the process, so that the various pros and cons were discussed both from an architectural and preservation point of view. This created a very positive relationship of trust, in which all parties were satisfied with the result, despite the surprising intervention and solutions.

p. 203: Main staircase linking all parts of the building together. A new roof over the former backyard creates an entirely new interior square, the "winter garden". Opposite: View looking up to the stairs leading into the building. Photos on pp. 202–205 by Bosheng Gan. Photos on pp. 206–209 by Lars Petter Pettersen.

第203页：连接建筑所有部分的主楼梯。原有后院上方的新屋顶创造了一个全新的室内广场“冬季花园”。对页：仰视通往大楼的楼梯。

新的文化中心森特拉伦是克里斯蒂安娜斯巴达银行的旧银行大楼（1899年竣工）和邻近的办公楼（1900年竣工）的二次开发利用，这两座大楼是斯巴达银行基金会于2007年决定改造的。2012年，奥斯陆工作室和KIMA建筑设计事务所赢得了一个开放的建筑竞赛。在这个竞赛中，他们提出的方案仅包括一套工作流程以及对它的描述，没有任何图纸。工作流程设想了如何设计一个新的文化中心，以及如何实施这样的重建工程。建筑的平面布局和使用方法在当时还是完全开放的。

在旧建筑改造过程中，项目重点被放在了建筑的原有特色上，并尽可能地对其加以再利用。因此，找出建筑中那些隐藏在地毯、石膏和天花板等表层之下的特点并决定哪些可以被保留，成了一项令人兴奋的工作。在这一过程中，许多装满残片的集装箱被从现场运出。为了重新开放这座原银行大楼，项目对受到保护的立面和内部景观做了一些结构上的修改，让大楼的氛围变得更有魅力和友好。由于旧建筑从未被设计成相互连接的样式，所以最大的新变化就是将这两座建筑连接成一个新的整体。项目新建了一个入口与奥夫尔城堡街（Øvre Slottsgate）相连，并增加了一部电梯和一个楼梯，将建筑的各个部分连接在一起。原有后院上方的新屋顶创造了一个新的室内广场“冬季花园”，它已成为游客和使用者亲近自然的打卡点。

在改造过程中，尽可能少地增加新元素非常重要。最终方案只增加了以防火和隔音用的隔墙、通用设计元素、吸声结构等为代表的基础技术设施。大部分旧的饰面表面得以保留了下来，仅有一些特殊的房间被修复且被大胆地变换了用途，但依旧保留了它们原有的风格。这种粗犷与典雅的对比，给这座建筑带来了一种独特的氛围。

改造过程中，为了让相关项目参与者们更好地理解改造思路，项目还建造了一个大型建筑工作模型。在模型中，人们对新的流线、装饰和设计的不同解决方案进行了测试和讨论，参与者可以设想各种房间的新用途，以及这些房间之间应有的关系。

这个项目特别强调灵活性和功能性的解决方案。所有房间都是多功能的，可用于各种活动。同时，工作站的数量和大小都易于调整。所有的技术设备都裸露地安装在天花板上，这使得调整变得非常简单。这是一个经得起时间考验的项目。

这座建筑具有非常高的保存价值。项目初期便与奥斯陆市的市政文化遗产管理办公室建立了良好的对话关系，在建筑和遗产保护角度讨论了各种利弊。这种做法建立了一种非常积极的信任关系，尽管项目中存在一些令人惊讶的革新干预和解决措施，但各方对最终结果都感到很满意。

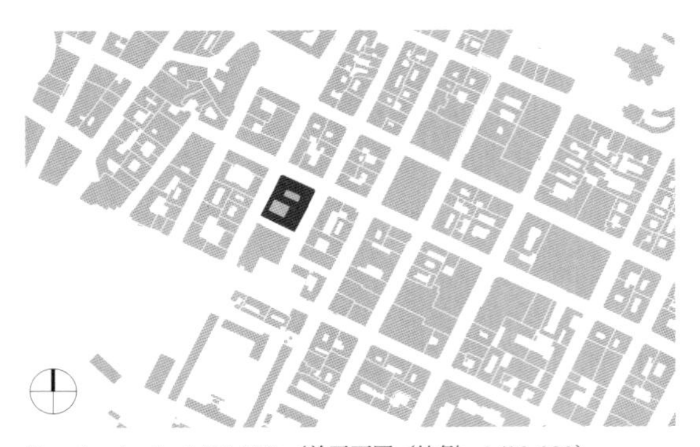
Site plan (scale: 1/10,000) ／总平面图（比例：1/10,000）

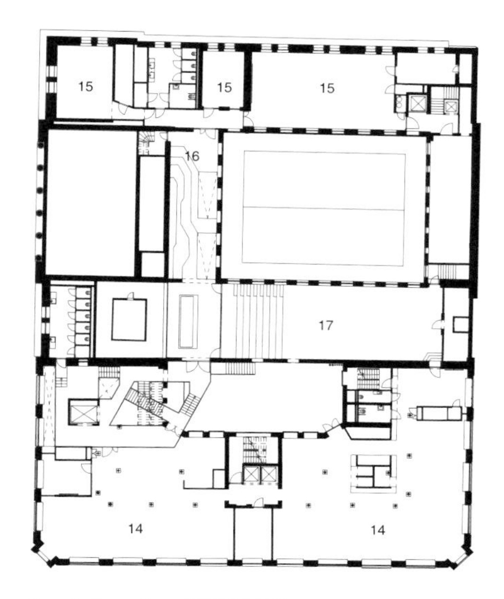

Fourth floor plan ／四层平面图

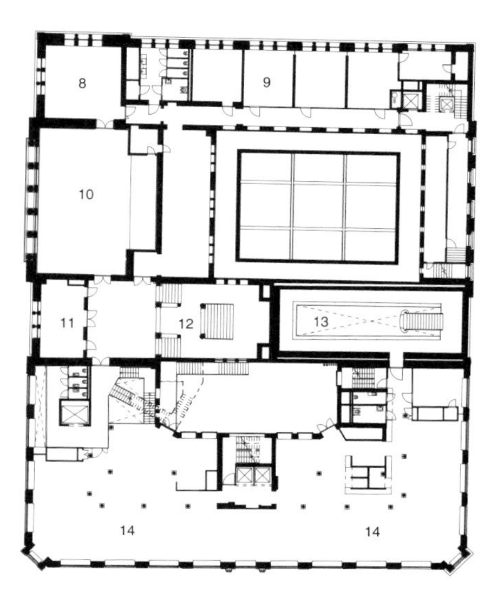

Third floor plan ／三层平面图

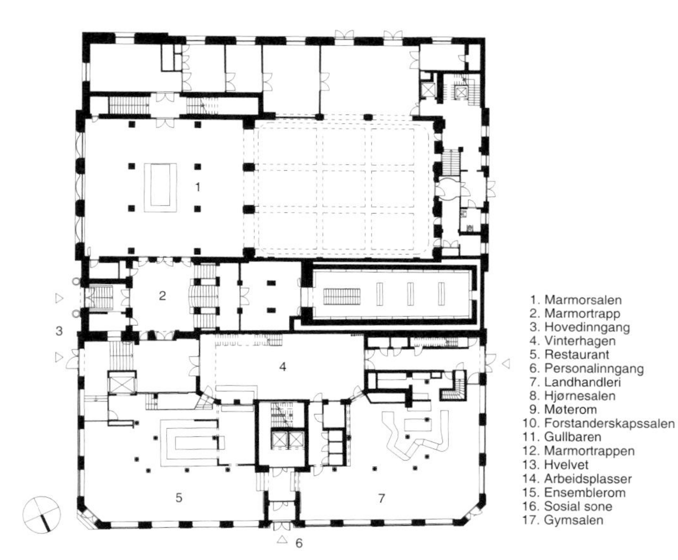

1. Marmorsalen	1.大理石大厅
2. Marmortrapp	2.大理石楼梯
3. Hovedinngang	3.主入口
4. Vinterhagen	4.冬季花园
5. Restaurant	5.餐厅
6. Personalinngang	6.员工入口
7. Landhandleri	7.展览室
8. Hjørnesalen	8.角厅
9. Møterom	9.会议室
10. Forstanderskapssalen	10.供给室
11. Gullbaren	11.会计室
12. Marmortrappen	12.大理石楼梯
13. Hvelvet	13.保险库
14. Arbeidsplasser	14.办公区
15. Ensemblerom	15.私人办公室
16. Sosial sone	16.公共活动区
17. Gymsalen	17.健身房

Ground floor plan (scale: 1/800) ／一层平面图（比例：1/800）

Opposite: View of the bank's facade.p. 208, from above: View of the gym on the above floor. View of the vault in the old bank building. View of the social zone. p. 209: View of the old bank's marble hall.

对页：银行的外立面。第208页，从上到下：上层体育馆；原银行大楼保险库；社交区域。第209页：原银行大理石大厅。

Atelier Oslo

The New Deichman Library

Oslo, Norway 2009–

奥斯陆工作室
新戴希曼图书馆
挪威，奥斯陆 2009-

The New Deichman Main library is the new public library for the city of Oslo. In 2008–2009 a design competition was launched for the New Deichman Library with adjacent areas in the so-called Deichman axis. The project Diagonale authored by architects Atelier Oslo and Lund Hagem was unanimously chosen as the winner.

The City Council has high ambitions for the library. The library should be at the forefront in the development of the modern public library system and a strong visitor attraction. It should be a cultural institution with a clear profile of high quality and with high accessibility, as well as an important source of knowledge and culture to a wide audience.

Connecting to the city

The building's form is a response to the surroundings – the building weaves into the city fabric of Bjørvika and respects the sight lines to the Opera House. The cantilevered areas solve the space requirements for the library while providing the building with a distinctive shape and a clear orientation towards the city and the library plaza in front. The interior spaces are organized around three diagonal light shafts (voids) that spread light and create contact between the different levels and functions of the library. From each of the three entrances, sightlines through these light shafts create contact between the interior of the library and the city, creating an interactive and inclusive building. The spatial organization creates a great variety of different spaces and places. The new Deichman will be a modern library built to a human scale, where the human being is in the center, not the stack of books. The ambition is that the new library will be the new main meeting place for the citizens of Oslo.

The facade diffuses sunlight, giving a calm feeling to the interior. At night, the building will glow and change appearance, communicating the various activities and events inside the library to the surrounding city fabric.

Energy efficient

The new library will use less energy than any of the new cultural buildings in Norway. This is achieved with a compact building where facade and ventilation systems are optimized. Ventilation air runs in the floors and air aggregates are distributed in a decentralized system to use less energy to move the air. Cold water runs inside the concrete slabs that are exposed in the ceiling to cool the building. The structure of the facade breaks ground by using glass fiber composite (GFRP) elements in the construction. This allows for a facade structure that minimize cold bridges at the same time as the configuration gives an even distribution of daylight into the library spaces. The facade is a further development of the competition entry whose design ambition was to create a unified and calm skin for the building. For that, the design is based on the organization of translucent and transparent areas that will make the building glow at night. Towards the ground the library has a completely transparent glass facade that opens the library towards the surrounding city and makes the ground floor an extension of the surrounding public spaces.

INFO

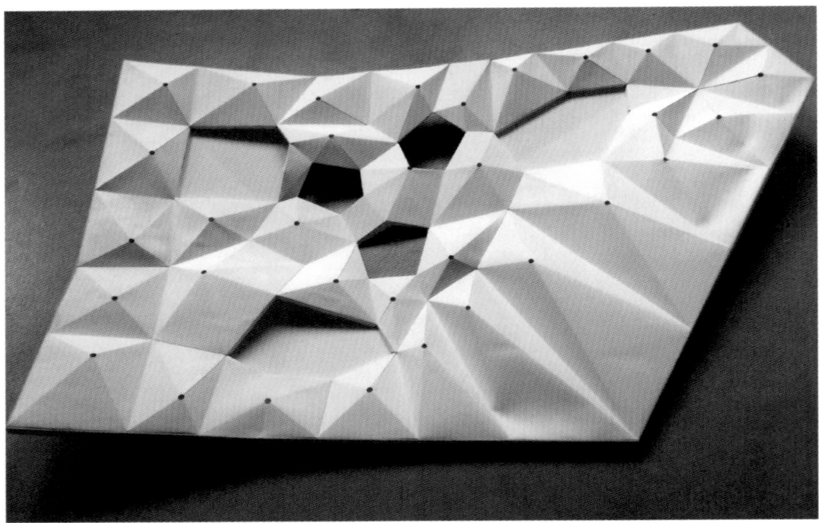

p. 211:Render view of library plaza. All photos on pp. 210–217 courtesy of Architect unless otherwise noted. This page, above: Model of the three diagonal light shafts. This page, below: Ceiling model.

第211页：图书馆广场效果图。本页，上：三个采光井的模型；本页，下：天花板模型。

这座新戴希曼图书馆将成为奥斯陆市新的城市图书馆。从2008年到2009年，在被称为“戴希曼轴”的附近的区域，一个新戴希曼图书馆的设计竞赛开始了。由奥斯陆工作室和伦德·哈格姆共同设计的“对角线”方案被一致评选为优胜者。

市政府对这座新图书馆抱有很高的期待。它应走在现代公共图书馆系统发展的前沿，并对游客有强烈的吸引力。图书馆也应是一个形象鲜明、高质量且可达性强的文化机构，并成为广大市民学习了解知识和文化的重要来源。

与城市相连

这座建筑的外形是对周围环境的一种回应，它融入了比约维卡地区的城市结构，但并没有破坏人们欣赏歌剧院的视线。悬空结构解决了图书馆的空间需求问题，同时使建筑具有独特的形状，赋予了建筑面向城市以及前方图书馆广场的明确朝向。内部空间是围绕着三个采光井组织起来的，这些采光井不仅可以扩散光线，还在图书馆的不同楼层和功能之间建立联系。视线能够通过三个入口，穿过采光井在图书馆内部和城市之间建立起联系，营造出一个互动包容的建筑。这样的空间结构创造出了各种各样的空间和场所。新戴希曼图书馆将是一个依照人体工程学建造的现代化图书馆，在这里，中心是人，而不是堆积如山的书籍。这个项目的愿景是让新图书馆成为奥斯陆市民新的主要集会场所。

立面能够缓和日光照射，为室内营造一种平静的氛围。到了晚上，建筑会发光并改变外观，将图书馆内举行的各种活动与仪式传达给周围的城市空间。

能源效率

新图书馆将比挪威任何一座新的文化建筑都节能。这是通过最大限度优化立面和通风系统、让建筑变得更加紧凑来实现的。在各楼层中流动的空气和集中的空气被散布到一个去中心化的系统中完成，这样便减少了输送空气所消耗的能量。天花板中裸露的混凝土楼板中间有冷水流过，用来帮助建筑降温。建筑的玻璃幕墙在施工时采用了玻璃纤维（GFRP）构件，这使得立面结构能够最大限度地减少冷桥效应，同时使日光均匀地分布到图书馆空间中。这个立面是在竞赛方案的基础上进行了进一步展开，其设计目标是为建筑创造一个和谐、稳重的表皮。为此，立面的设计基于对半透明和透明区域的组织，使得建筑能够在夜间发光。图书馆的玻璃立面在接近地面的地方会变得完全透明，使图书馆向周围的城市开放，并使一层成为周围公共空间的延伸。

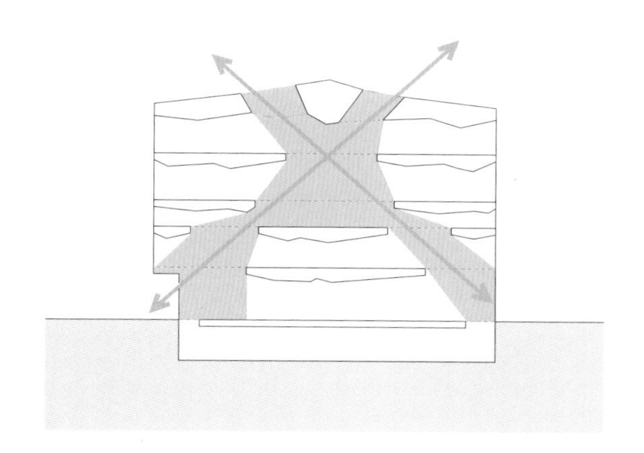

Concept section ／概念剖面示意图

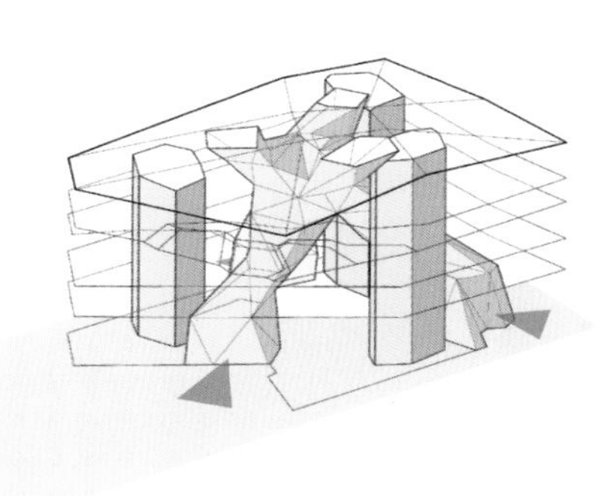

DEICHMANSAMLINGEN
戴希曼系列

AUDITORIUM / KINO
礼堂/放映厅

SERVERING
服务区

PUBLIKUMSAREALER
公共区域

ADMINISTRASJON / DRIFT
管理部

VARELEVERING
订阅区

MAGASINER
杂志区

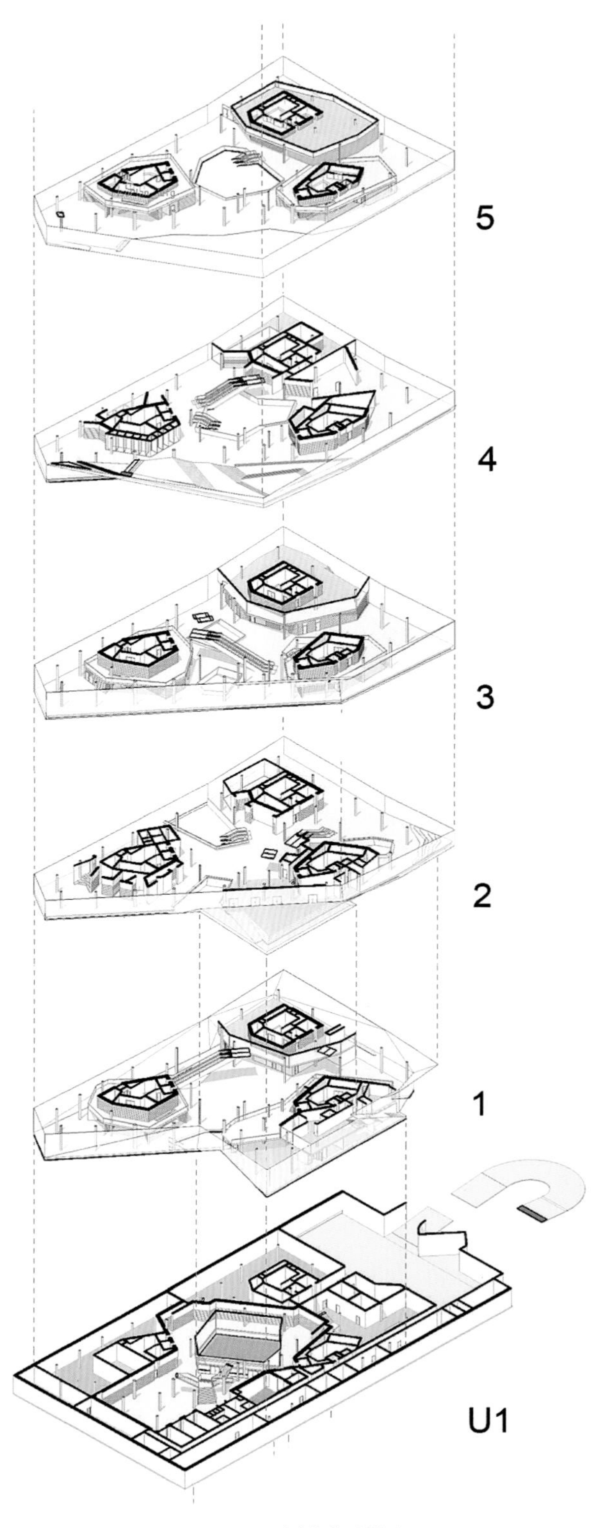

Diagram of overview functions ／功能分区总图

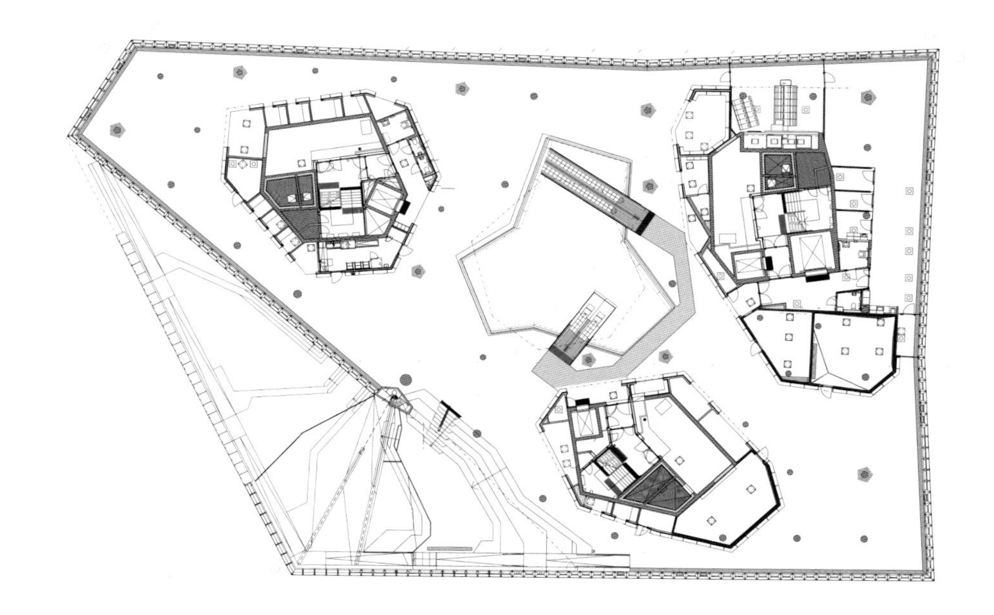

Sixth floor plan ／六层平面图

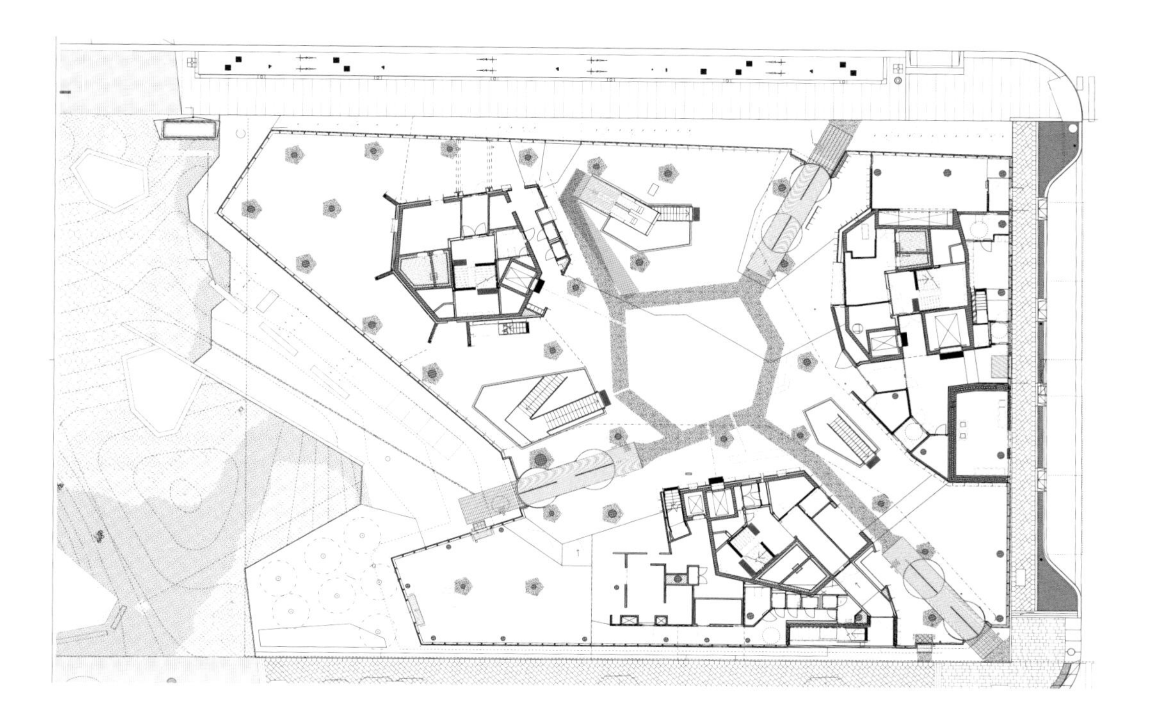

Ground floor plan (scale: 1/650) ／一层平面图（比例：1/650）

Credits and Data

Project title: The New Deichman Library
Client: The municipality of Oslo
Program: Main Public Library
Location: Oslo, Norway
Design: 2009–
Status: Under construction – estimated opening 2020
Architect: Atelier Oslo and Lund Hagem
Landscape architect: Agence Ter / SLA
Gross area: App. 20,000 m²

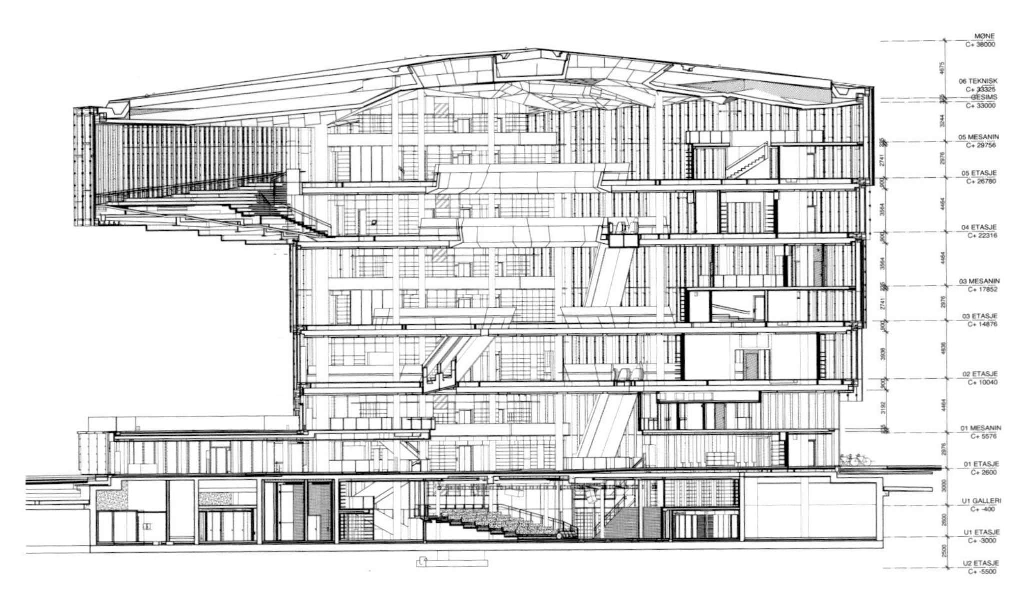

Long section (scale: 1/600) ／纵向剖面图（比例：1/600）

p. 214, above: Aerial render view of Library. p. 214, middle and below: View of the building under construction. Opposite, above: The New National Opera House in Oslo by Snøhetta. View from the interior still under construction. Opposite, below: View of the 3 diagonal light shafts spreading light into the library space. Photos on p.214 middle, p.217 by Bosheng Gan. Photo on p. 214 below by Thomas Rønhovde.

第214页，上：图书馆鸟瞰效果图；中和下：建筑施工期间的景象。对页，上：斯诺赫塔建筑事务所为奥斯陆设计的新国家歌剧院。当时建设中的图书馆的室内；对页，下：三个采光井将光传播到图书馆内部空间。

IAM
致敬20~21世纪传奇建筑家
全解建筑世界里的光影挑战
基本信息：开本16开/尺寸215mm×280mm/语言 内文全中文，索引中日英三语/页码3496页（以最终印刷版为准）/结构
马卫东/执行编译 安藤忠雄全集编辑部 / 出版发行 中国建筑工业出版社 / 主要作（译）者 安藤忠雄 肯尼斯·弗兰姆普顿

Spotlight:

Xinhua Culture & Creativity Light Space / Pearl Art Museum

Tadao Ando Architect & Associates

特别收录：

新华文创·光的空间·明珠美术馆

安藤忠雄建筑研究所

At the end of 2017, "Pear Art Museum/ Xinhua Bookstore" was finally completed. This is an architectural work in which we completely redesigned a portion of a Shanghai commercial complex known as "Xinhua Red Star Landmark" while the construction of the building was already in progress.

The final decision to accept the projet was due in part to my spirit to undertake challenges and face the unknown but also how deeply the owners' passionate ideas and devotion to create a "cultural landmark in Shanghai" had touched me.

Bookstores or libraries, which live through books, should be an important carrier of knowledge in any era. Such is art. It exists in a diverse and complex society. Its existence remains in question but, to understand the spirit and value of human existence in this era, art should be the most appropriate medium. Excellent art works can directly reach people's hearts and sometimes help us wake the sensibilities that lay sleeping.

"Pear Art Museum/ Xinhua Bookstore", as a venue of "books"and "art"for the creation of a new urban future, is an ideal urban facility that I once pictured in my inner heart. Now, when I look back, this special design condition of "building while thinking" is quite appropriate to a project.

As a result of rare design circumstances, we believe a unique space was born. We redesigned a portion of the top two stories with a bookstore on the lower level, a museum on the upper level, and a double-height egg-shaped space filled with bookshelves connecting the two. The egg-shaped space is known as "Oval Room" and is primarily utilized as a cafe and a multi-purpose space capable of hosting various events. The unique relationship between the upper and lower floor plays a central role in the internal spatial configuration of the architecture. Additionally, this internal egg-shaped space becomes evident in the exterior facade as well, elevating the building to an iconic architectural landmark.

The original idea of the "Light Space" is to breed new life, as birds incubate their eggs. I endeavored to express our hope for the future through such an impression and make a mark upon the Shanghai skyline. May this "egg"also give birth to a "cultural light" through the illumination of wisdom in books and arts.

I look forward to seeing the creations inspired by this "egg" that floats in the vault of heaven in Shanghai and may it vibrantly and continuously renew life and hope.

Text by Tadao Ando

Credits and Data

Project title: Xinhua Culture&Creativity Light Space/Pearl Art Museum
Location: Shanghai, China
Design: 2015.9-2016.11
Completion: 2016.8-2017.12
Structure: Reinforced Concrete, Steel
Fuction: Book shop(7F), Museum(8F)
Site area: 9,720 m^2
Total floor area: 3,975 m^2(7F 1,550 m^2; 8F 2,425 m^2)

pp.220–221: Oval Room in "Light Space" with star sky dome. Opposite, above: Overlooking the "Light Space"at night. Opposite, below: Hand drawing by Tadao Ando

第220-221页：拥有星空穹顶的“光的空间”心厅。对页，上：夜幕下，远眺“光的空间”；下：安藤忠雄手稿。

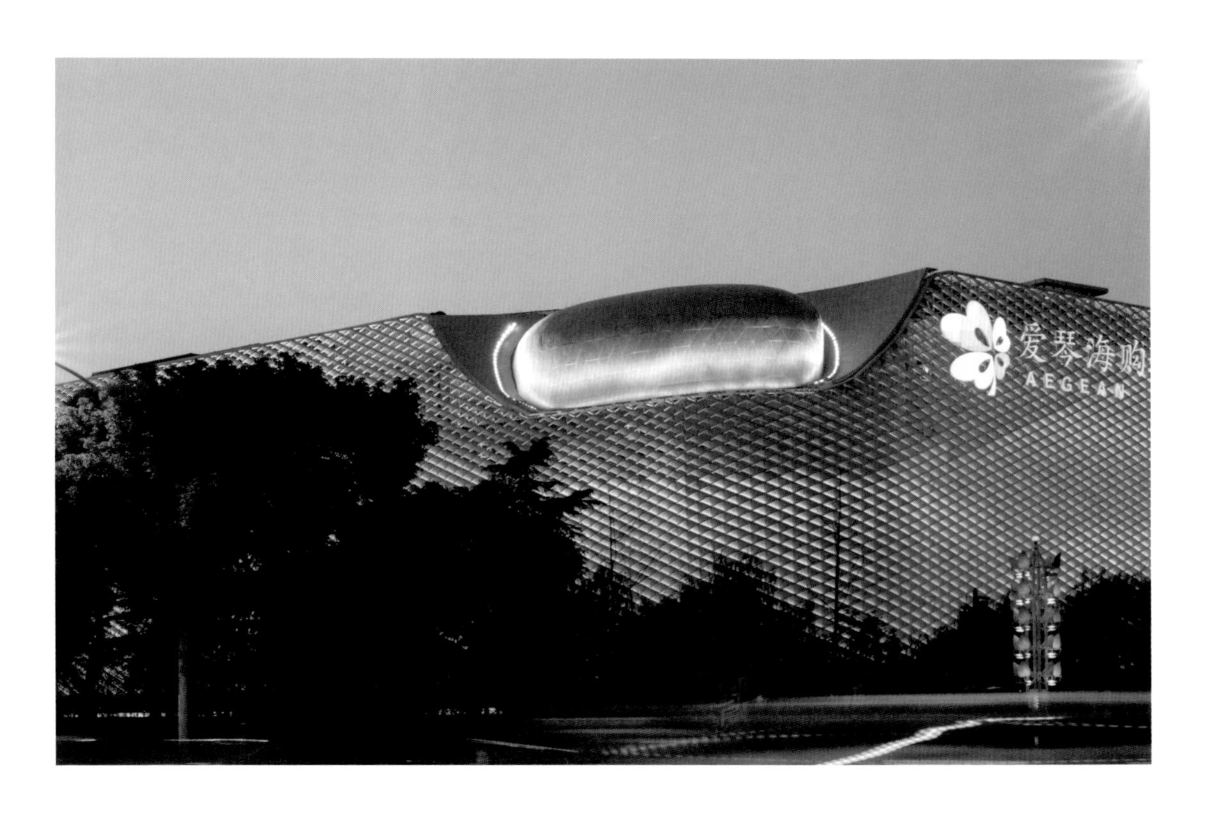
爱琴海购
AEGEAN

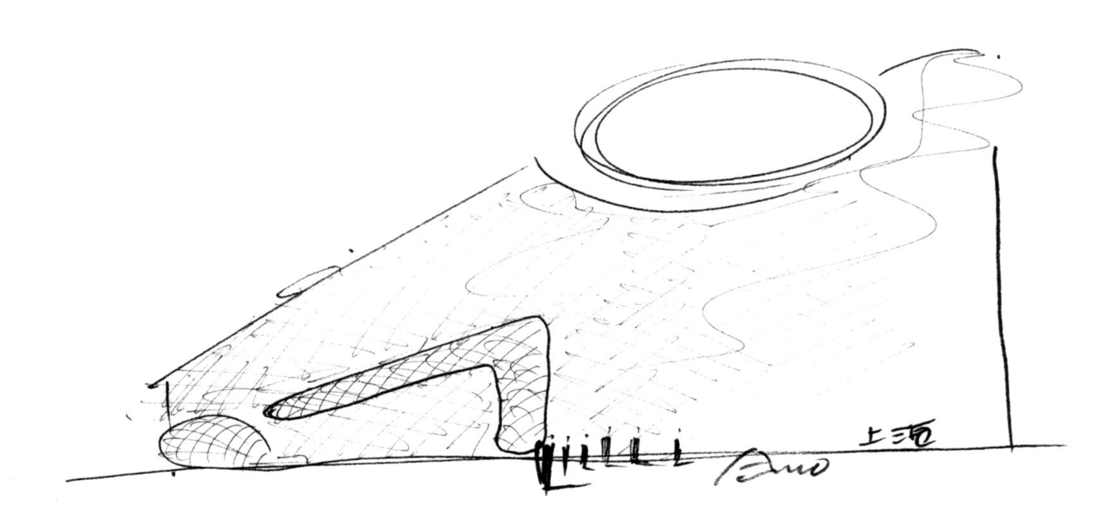
上海

2017 年终，光的空间“明珠美术馆 / 新华书店”竣工落成。这是我们对一座业已在建的商业综合体（新华 · 红星国际广场）的一部分空间内外进行重新设计后所诞生的一座“建筑”。

后来之所以接受这个项目的委托，其中既有自己内心对未知之事的挑战欲的躁动，但更多还是因为业主，他们倾注于这个建筑中的想法，他们的梦想——“成为上海的一个文化地标”，这些都深深打动了我。

以书为营生的书店或图书馆，它们无论处于哪个时代，都应是知识的重要平台载体。艺术也是，艺术存在于多元化、复杂化的社会中，它的存在意义虽还是个疑问句，但要了解生存于这个时代的人类精神、价值，艺术应该是最恰当的一种媒介吧。优秀的艺术作品，它们会直扣人心，有时还可以帮助我们唤醒内在沉睡的感性。

以艺术为营生的美术馆，不应囿于规章制度的框架约束，而应更真挚地作为公共之用。这个为创造新都市未来而打造的“书”与“艺术”的场所——“明珠美术馆 / 新华书店”，正是一座我曾经在自己内心勾勒描绘的城市设施。如今再想起来，这种“边建造边思考”的特殊的设计条件，说是相宜于项目的，也未尝不可啊。

系于结果，我们希望可以营造一座与设计条件之独特相契、极富个性的“建筑”。设计以建筑上层部分的其中两层为对象，下层设书店，上层为美术馆，中央置内壁被书架墙通围的卵形挑空空间。该卵形空间“心厅”，日常可作与书店一体的咖啡吧，生活日则可用为多功能厅，它既衔接起上下层，成为建筑内部结构的核心，其独特的空间形体又径直化为建筑整体外观的一部分，将会提升整座建筑的地标性。

鸟翼觳卵，鸟儿怀抱着一颗“蛋”的样子，就是我寄寓在“光的空间”中的创想原点。我希望通过这样的一种设计印象与上海这座跃跃之城的天际线一气承转，以表达我们对未来的希望。而从这颗“蛋”中即将会诞生的，即由书籍与艺术共酿的“文化之光”。

我期待着这颗浮于上海天穹的创造之“卵”，作为一个新的城市创造据点，强而有力地活着，并不断孵化。

安藤忠雄 / 文

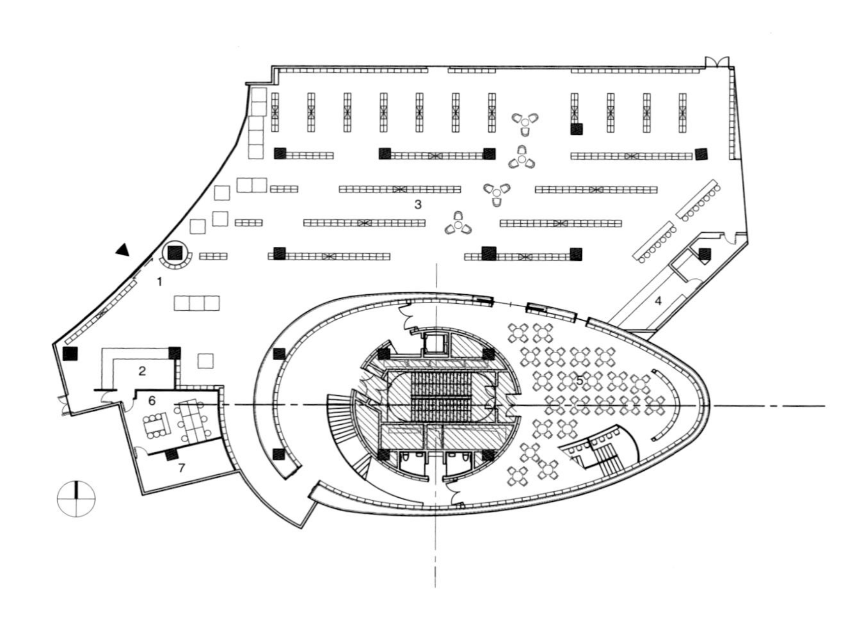

Seventh floor plan (scale: 1/600) ／七层平面图（比例：1/600）

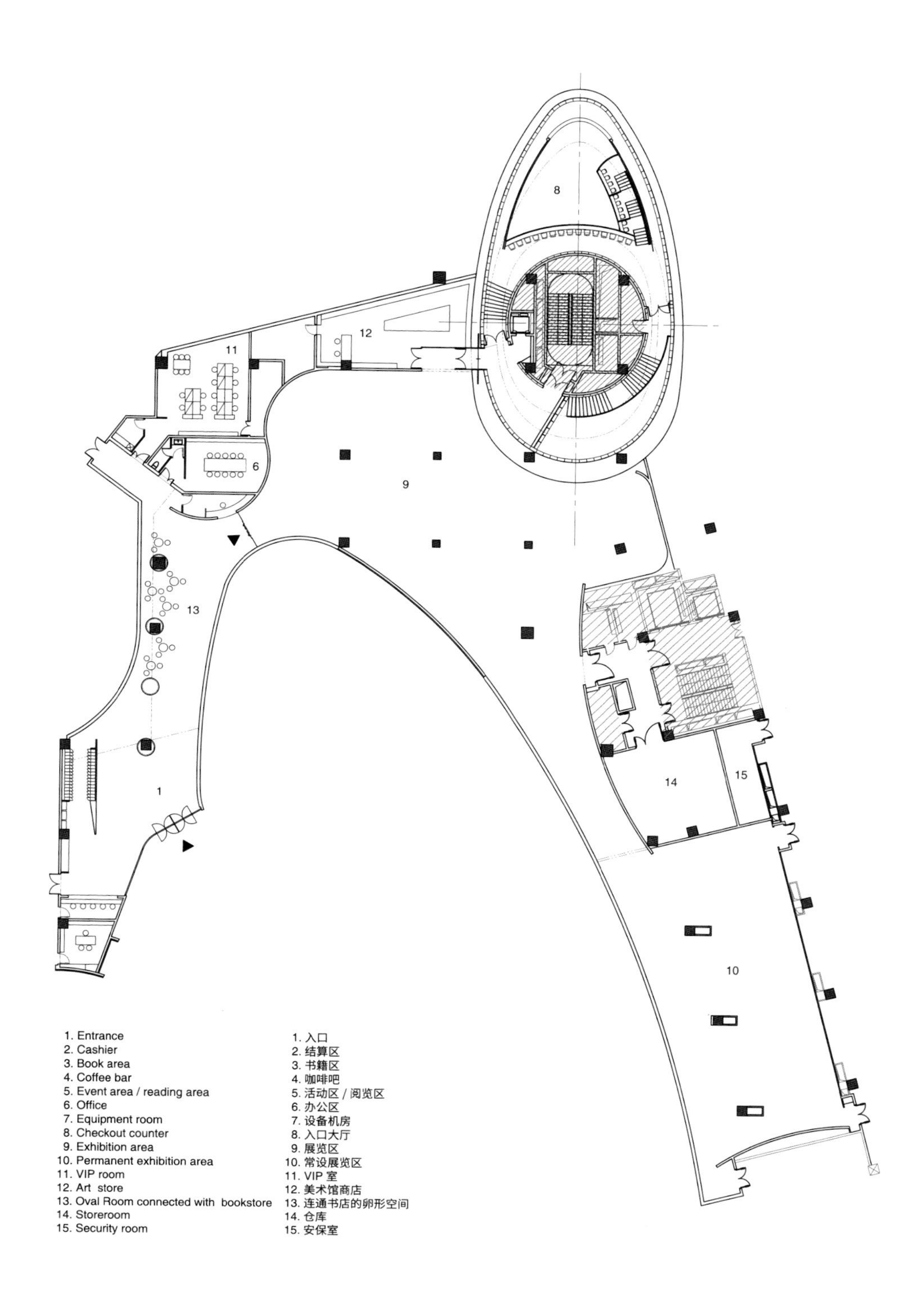

Eight floor plan／八层平面图

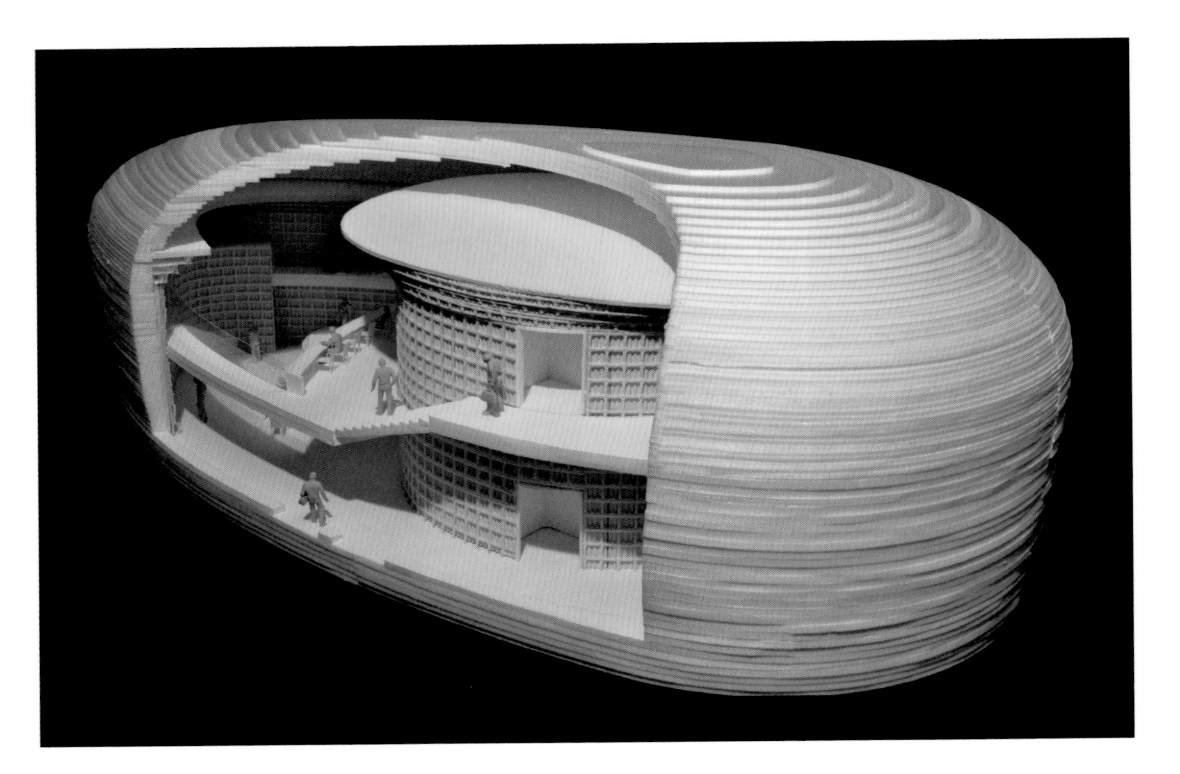

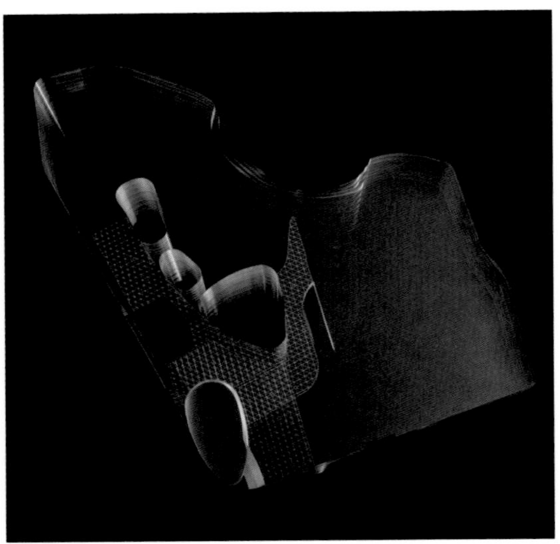

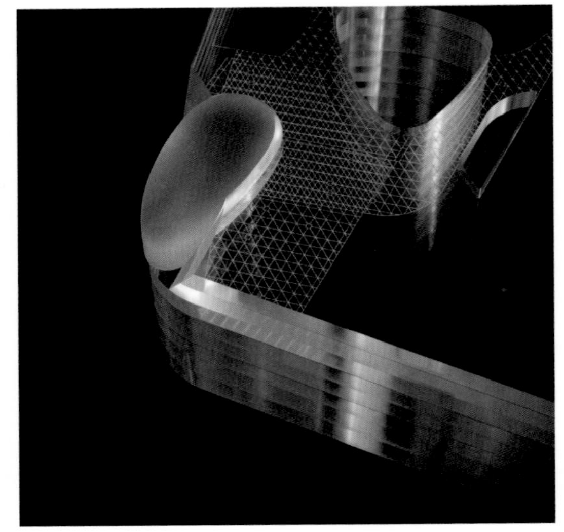

This page, above: Diagram of the oval space model. This page, below: Final scheme model diagram. Images by Tadao Ando Architect & Associates. Opposite: Ring bookshelf inside the Oval Room.

本页，上：卵形空间模型图；本页，下：最终方案模型图。对页：卵形空间内部的环形书架。

7TH HEAVEN

This page, above: Well-proportioned bookshelves inside the bookstore. This page, below: Fair-faced concrete stairs and overhangs in the Oval Room. Opposite: In the Pearl Art Museum, you can see the photographic works in distance through the circular and square geometric openings. pp. 230-231: Bookshelf array with embedded windows for space penetration power. All images on pp. 220–231 by CHEN Hao except the specified.

本页，上：书店内错落有致的书架排布；本页，下：心厅内清水混凝土楼梯和挑檐。对页：明珠美术馆内，透过圆形方形几何开口，看远处的摄影作品。第 230-231 页：具有空间穿透力量的嵌窗洞式书架阵列。

只有医生
知道！
睡眠
E2-9-6
轻松入门
插花
实用养花
实用养花

E2-11-6
图解人体大百科
E2-11-5
28天
内分泌平衡书
E2-11-4
图解舌诊
经络穴位
按摩大全
E2-11-3
国医大师的
养生茶
一本会卖萌的中医书
柔软的身心